UNIX

Computer Telephony

The Complete Guide

by John W. Kincaide
Scheduled Solutions Inc.

Flatiron Publishing, Inc.

New York

NY THE COMPLETE GUIDE

Flatiron Publishing, Inc. Book
Copyright © 1997 John W. Kincaide, Canada
Published by Flatiron Publishing, Inc.

Flatiron Publishing, Inc., New York.
12 West 21 Street
New York, NY 1010
Phone 212-691-8215 Fax 212-691-1191
1-800-999-0345
1-800-LIBRARY
www.flatironpublishing.com
PUBLISHER Harry Newton: Harry_Newton@msn.com

AUTHOR
John Kincaide
President
Scheduled Solutions Inc.
6 Glen Edyth Drive
Toronto, Ontario
Canada
M4V 2V7
416-929-0440 Phone
416-929-7927 Fax
johnk@scheduledsolutions.on.ca
www.scheduledsolutions.on.ca

ISBN # 1-57820-013-X

First Edition and First Printing, July, 1997
Manufactured in the United States of America
Printed at Bookcrafters, Chelsea, Michigan

Table of Contents

List of figures

List of Tables

Notation used in this book

The use of the notation below is to aid the reader in understanding what kind of speech or special reference is being made within the body of the text of this book. Although not used heavily, it is important to highlight this at this time.

Notation	Example use of Notation
Text enclosed by "< >" indicates recorded speech being played back by an computer telephony application.	For example <"two thousand, eight hundred widgets>"
Text enclosed by "{ }" indicates synthesized speech using text-to-speech being played back by an computer telephony application.	For example, a sales manager calls into a voice mail system, which integrates email for his messages. Because the system has text-to-speech capabilities it plays back the email Subject Lines over the phone of the unread email before the mailbox subscriber decides to fax this email message to his hotel. While reviewing his email the system would report an email subject line converted to synthesized speech as "{Weekly sales report from Charlene Townsley}"
All Monetary values are expressed in US Dollars, unless otherwise noted.	$1.00 ($1.37CDN) shows that 1 United States Dollar is equivalent to $1.37 Canadian Dollar
All Units of Measure are expressed using the metric system. Equivalent United States unit of measure is documented in brackets.	10.67m (35 ft)

Notation	Example use of Notation
manday	This is used in reference to a single 8 hour working day. The reference to "man" in "manday" is not meant to be gender biased. It is also refers equally to both male and female workers.

Foreword

Harry Newton, the world's leading publisher and noted speaker of computer telephony integration technology, had just completed an informative and entertaining speech on the importance of the convergence of computers and telephony at the outdoor quarry amphitheater at the University of California at Santa Cruz. It was late August, and he was the keynote speaker of the last day of the SCO Forum95. After finishing the speech, a small group including colleagues, SCO staff and myself gathered around Harry, standing under a pine tree at the back of the amphitheater. Harry showed us the cover of his next issue of Computer Telephony Magazine - the famous "Teaching Pigs to Fly" issue, in which he discussed the closed architectures of PBX manufacturers.

Harry turned to the assembled group and asked "Why UNIX for computer telephony?" The conversation turned to the ability of the UNIX platform compared to other operating systems of the time. After I was temporarily distracted from the conversation, I heard Harry ask, "Would someone like to write *something* on why UNIX is a good computer telephony platform?" There was a moment of silence by all who were now looking at Harry. There was only a slight hesitation on my part, momentarily wondering why everyone was so stone faced silent. I immediately thought I could write something on this subject. So I volunteered by literally stepping forward towards Harry saying, "I will," thinking I would really like to write an *article* for his magazine. What I had not realized was that Harry had asked "Would someone like to write a ***BOOK*** on UNIX based computer telephony integration?" After having realized my 'mistake' - and it was my mistake alone - I kept my commitment to Harry, whose support throughout the project has been absolutely wonderful. My computer telephony adventure had taken a much different path than I had ever expected it would - all very much for the better.

What you now have in your hands is the result of my efforts and the cooperation of many others, which began as a momentarily lapse of attention, culminating in a book that is now over *462 pages in length, consisting of an estimated 108,714 words.* My adventure has led me to talk with and learn from many industry leaders from as far as Hong Kong, Australia, Europe, and many points within the USA and Canada.

This book's focus is on the business manager who needs to understand the return of investment and implementation of enterprise-strength computer telephony integration, as it pertains to the UNIX platform. The book takes the reader from simple concepts such as how a telephone works, right through to how to create and implement a multi-site, international computer telephony integrated call center environment.

To support the business manager's understanding, I review 14 case histories that use a variety of computer telephony applications. They include Interactive Voice Response, Call Centers, Screen Pops, Fax Servers, PC Switch Technology and other types of computer telephony applications. They are summarized below in Vital Statistics Grand Summary on page xx, showing strong return on investment justification for UNIX enterprise-strength computer telephony technology. As well, there are discussions of the "Ripple Effect" return on investment. When Computer Telephony applications are installed, they provide a benefit for the companies that install the technology, but also provide a cost benefit for their customers. This is the ripple effect. It is synonymous with the proverbial dropping of a pebble in a still pond.

The book also contains a comprehensive enterprise multiple site request for proposal for a fictitious international biotechnology company. The objective of this section is to provide guidelines for drafting such a document. An innovative solution to the request for proposal is provided by a leading computer telephony integrator.

Many industry leaders have contributed their wisdom, especially **DOUG MICHELS**, Vice-President & Co-Founder of **SCO** within the Preface on page xxiii. This is a **MUST READ.** I thank them all very much for their contributions.

Take that first step to a rewarding future! Just like that first step I took in the quarry in August of 1995. Your adventure has begun! It is my sincere hope that this book will guide you in your computer telephony adventure. If you need some advice along the way, please do not hesitate to contact me at the email address or phone number below.

July 12, 1997

John W. Kincaide
President
Scheduled Solutions Inc.
6 Glen Edyth Drive
Toronto, Ontario, Canada
Tel: 416-929-0440
Fax: 416-929-7927
johnk@scheduledsolutions.on.ca
www.scheduledsolutions.on.ca

Vital Statistics Grand Summary Table[1]

Vital Statistics Grand Summary of 14 Case Histories		
Cost	*Statistics*	*# of Cases Noted*
Weeks of labor to install system	83	10
Initial System Cost	$1,270,000	6
Est. Ongoing support Cost	$15,000	1
System Size		
# Ports - Total/# Fax.	6,776	11
# Users	1,114,732	12
# Calls/month	8,445,617	13
Computer	Intel Pentium, Sun Sparc, RS/6000 and many other platforms.	14
Operating System	SCO UNIX Operating Systems, Sun Solaris, IBM AIX, Silicon Graphics Irix, Data General Aviion, QNX	14
Industry	*Various major Industries*	
Estimated Return on Investment	unable to summarize at this time	
Cost Savings		
Manpower equivalency per year - in man years	241	10
Estimated Manpower savings per year	$11,351,600	4
Reduction in telephone costs/year	$2,294,600	3
Earnings		
Generated new income/year	unable to summarize at this time	
Ripple Effect		
Estimated cost Saving to customer base expressed in man years	15	7
Vendor	*Various UNIX based vendors*	
Product	Various products	
End User	14 different end users	

Vital Statistics Grand Summary of 14 Case Histories

Cost	Statistics	# of Cases Noted
Computer Telephony Technologies Deployed	Number of instances of deployed technology utilization within Case Histories	
Call Conferencing	1	
Call Control via Computer Workstation	1	
Client-Server base Computer Telephony Integration	4	
Fax-on-Demand	3	
Fax Server	4	
Interactive Voice Response	7	
Internet based Fax Solutions	1	
Intranet based Voice transmission	1	
ISDN distributed call control	1	
Large vocabulary speaker independent recognition	1	
Middleware	1	
PC Switch	3	
Worldwide remote CTI system management utilizing UNIX standards	2	
Computer Telephony Integration Screen Pop	3	
Text-to-Speech	1	
Voice Messaging	1	

Table 1-1 Vital Statistics Grand Summary of 14 Case Histories

---

[1] Dear Reader: Throughout this book, each case history is summarized at its conclusion by a "Vital Statistics" table. There are 14 case histories in this edition. The numbers you see in this table are summations of all the facts across all the case histories.

However, as Prime Minister Disraeli, of the United Kingdom during the reign of Queen Victoria in the 19th century said this about statistics: "There are lies, damnable lies, and then there are statistics!". There is a lot of truth to his statement. First of all, the information collected in the case histories is second hand information through interviews of the participants in the installations. In other words, this information was not directly measured by the author. As well, it is important that the case histories are from companies who volunteered the information. The sample is by no means a scientific one. For example, it was not drawn from a random sample. Moreover it is important not to be tempted to average or do other advance calculations within this grand summary table, since it is only an addition calculation of those case histories that did provide some information. There are data points where no information was volunteered for confidentiality reasons. So this could really bias the results of an statistic such as the "average return on investment" etc. Some of the information is derived by assumption from other facts that are provided.

The column within Vital Statistics Grand Summary named "# of Cases Noted" indicates the number of case histories that reported that statistic. This is given so that one can "weight" or see the applicability of the sum in the previous column. The section "Computer Telephony Technologies Deployed" within the table does not have this last column filled in because it would be redundant.

So as not to call the purpose of the Vital Statistics tables into question, it is important to note that they are provided as a means to help the reader understand the potential impact of UNIX based computer telephony technology, rather than provide a definitive, and demonstrable result that one would expect within your company. You should at least be able to use the knowledge within the case histories as a basis for investigation within your company of how to measure the impact of computer telephony integration. This will help you to obtain the desired results on your company's bottom line. The Vital Statistics tables and the discussion about them is meant to provoke you to set up your own means of dependable measure of success.

Preface

By Doug Michels, Vice-President & Co-Founder of SCO[3]

SCO is the world's leading world's leading supplier

of UNIX server operating systems

My friend John Kincaide has written more than 400 pages about computers, telephony and UNIX systems. John is a real pro and has avoided unnecessary jargon in favor of good ol' standard English, so by the time you've read, say, half the book you'll understand UNIX® systems and

Figure 1-1 Doug Michels speaking at SCO Forum. SCO is the world's leading supplier of UNIX server operating systems. Copyright 1996 SCO, used with special permission.

Computer Telephony. If it's your business to understand enterprise-strength computer telephony, then you've come to the right place.

What this book is really about -- and I think John agrees with me on this one -- is running a more efficient, more customer-focused, more trouble-free business in an era when all three qualities are needed for success. And I don't believe there is any operating system that compares with UNIX systems in helping to build such a business.

Given that John has targeted non-technical managers as a key audience for this book, let me offer a little history. The UNIX operating system and telephony work together so well because the UNIX operating system was designed by the telephone company, back in the days when that meant AT&T and only AT&T. Ma Bell built the UNIX operating system to control both telephone equipment and run its business applications. At the time, AT&T was prohibited from entering the computer business so it gave the operating system away, mostly to colleges.

So several generations of young programmers have learned UNIX systems and helped improve upon it. It's a process which continues to this day. My company's contribution has been to popularize UNIX operating system on Intel-based personal computers. People who want UNIX operating system to run on their Pentium machines come to SCO. Alternatively, they can invest in larger UNIX servers from Sun, HP, and IBM. SCO has been very successful in providing many core technologies such as X-Windows. Moreover, SCO has been very successful in computer telephony systems, as evident within this book.

One great thing about UNIX systems is that it's been around a relatively long time -- at least by computer industry standards -- and is used to run some of the biggest and most complex business applications. Microsoft is making a lot of noise about its Windows NT operating system, a newcomer compared to UNIX systems. Someday NT may offer the reliability and scaleability as well as the wealth of talented programmers and administrators which already exist in UNIX systems. But that's years in the future and your business needs a hot telephony solution right now.

If that's the case, I don't see what choice you really have. The UNIX operating system is the only operating system that really meets your needs -- a point amply demonstrated in this book. Even if you don't buy SCO UNIX based computer telephony applications, please go to HP, Sun, IBM, or any of several other UNIX vendors and buy theirs. The UNIX operating system is far better suited for enterprise-strength computer telephony applications than Windows NT.

In this book you'll find business solutions you probably didn't know existed. Certainly I did. John has searched a number of industries to find the very best implementations of UNIX system based computer telephony applications. This is information which has never been gathered together in one place ever before and which has taken years to assemble. If you follow his advice I'm sure you will be well-served.

Finally, this is also a book about creating heroes in business. The right computer telephony system can literally make a business. Not surprising, the wrong system can help wipe a business out. Reading this book will arm you to build the right system and avoid the wrong ones. Act on this information and you'll be a hero-in-the-making.

I wish you good luck in your UNIX computer telephony adventure.

---

[2] Portions Copyright 1997, The Santa Cruz Operation, Inc.

[3] About SCO

SCO is the world's leading supplier of UNIX server operating systems, and a leading provider of client-integration software that integrates Windows PCs and other clients with UNIX servers from all major vendors. SCO is committed to bringing the Internet Way of Computing™ to business-critical environments of all sizes. SCO Business Critical Servers run the critical, day-to-day operations of large branch organizations in retail, finance, telecom, and government, as well as corporate departments and small to medium-sized businesses of every kind. SCO sells and supports its products through a worldwide network of distributors, resellers, systems integrators, and OEMs.

For more information, see SCO's WWW home page at http://www.sco.com/.

For more information SCO Telecommunications,
http://www.sco.com/solutions/telephony/index.shtml

Acknowledgments

There is not enough praise to be said for the many companies that have helped by providing their success stories. This includes all the manufacturers as well as their dealers and ultimately their customers - the end users. MANY THANKS TO ALL WHO PARTICIPATED!

A special mention for the SCO Telecoms Team, such as Gordon Jago, Tina Stewart, and Doug Warrilow for helping with ideas, support and sources of information for the book. My great appreciation goes to Doug Michels, Vice-President Technology & Co-Founder of SCO, for his support and for his preface.

A special credit goes to Trevor Strudley, Assistant to the CEO of SCO for technically reviewing the accuracy of this book. His efforts are deeply appreciated.

A special thanks also goes to other vendors such as Julia Klein at IBM Canada, and Steve Cawn of IBM USA as well as other UNIX manufacturers. For a complete list of vendors please see Appendix A.

A special thanks to Karen Richards and Kathleen Pacyna at Sun Microsystems, Inc. in their support for the book.

A special thanks to Cyril Kincaide, M.D. for taking the time to grammatically edit these pages. His patience and sincere desire, help and encouragement will always be remembered and treasured.

To my friend from Australia, David Bilbow, for all his words of humorous encouragement during my late evenings (his early mornings) while crafting this manuscript. To Sylvia and Les Kovaks for their friendship. A special mention to Lois Loretta, Bill Laidlaw and Wendy Strong.

ACKNOWLEDGMENTS

Without Doru Partila's wondrous teaching efforts and technical knowledge of Microsoft Office technology and especially Microsoft Word 7.0, the presentation of the words, and graphics would have taken much longer and would have been a much more daunting task. Also credit goes to Peter Oliver, of Peter Oliver & Company (Markham, Ontario) for their graphic art support and help, and suggestions for some of the artwork in the book. A thank-you is in order to Keith Penner portrait photography for the book cover.

The author wishes to thank Darlene and John Ridout for their exemplary help in providing encouragement, research, and support in the development of this book. Without their help, this work would not have been possible.

A special thank-you is in order to Michael Seebeck, President of CallStream Communications Inc., who provided the time needed from my duties at his company, CallStream Communications Inc. to work on this manuscript. Michael Seebeck provided the opportunity for my entry into the Computer Telephony marketplace.

Praise goes to Paul Brandon, of Brandon Interscience of Cupertino, California for granting permission to use and modification of an existing request for proposal. This document was expanded upon by myself to create the enterprise wide, multiple site international call center fictitious Minden Biotech Request for Proposal. A very special mention is in order for both Richard Crouch, President and Steve Dunne, Vice President of Sales and Marketing of Capri Systems Inc., of San Jose, California for their efforts and support of the Response to the Request for Proposal portion of this book. Their contribution, insight, and recommendations to help the readership can not be praised enough - Thank-you.

One very special acknowledgment needs to go to Harry Newton for providing me with the encouragement and opportunity to get this book completed. I've written it in ski chalet's when I should have been skiing with my daughter, in jets flying 10,660 meters over the USA & Canada, and as well on ocean going ferries between Nova Scotia and Prince Edward Island, and transcontinental trains, and of course in my office. This was truly an adventure and a tremendous challenge. It was more of a challenge than I ever thought it would be. Most importantly it has been fun.- Thanks Harry....

Finally, to my family, and especially to my daughter Joanna Louise Kincaide for the patience and understanding in allowing me the time to work on the book during the times we should have been together. I learned the following special phrase from her Dutch born mother, Helena, and I want to repeat it now for Joanna: "Ik hou zo veel van jou, snoepie".

John Kincaide
President
Scheduled Solutions Inc.
6 Glen Edyth Drive
Toronto, Ontario, Canada
johnk@scheduledsolutions.on.ca
July 12, 1997

The Computer Telephony Adventure

You too can Enjoy the Adventure and the Treasure

S tanding on the campus of the University of California at Santa Cruz's Cowell College, I took a moment to reflect on what I had seen a few weeks before at one of my client's offices. Cowell Campus is nestled amongst the redwoods in the Santa Cruz mountains about 600m overlooking Santa Cruz California and Monterey Bay. I was standing in the quadrangle looking towards the Pacific Ocean away in the distance with brilliant blue August sky above. The morning coastal fog was hugging the edge of the football field below, giving you the false impression that you could walk out on top of the fog cloud over Santa Cruz into the middle of the bay. I was attending the Santa Cruz Operation Inc. (SCO) Forum '92. SCO is a leading UNIX operating system developer and specializes in UNIX on the Intel platform. SCO is the largest volume vendor of UNIX server licenses shipped worldwide.

At the time I was working for a distributor/reseller who was selling SCO products. The client I was reflecting about was CallStream Communications Inc. I had seen at their offices the ability to quickly create an interactive voice response application. Through an application generator, I was shown how an interactive voice response application could front end an existing database application, turning the telephone into a data terminal by accepting input via Touch-Tone and then speaking out or faxing information. I thought this was very intriguing, and wondered how this technology could be applied to a very large number of existing

UNIX based applications that were based upon SCO UNIX. Order entry, order status, and a host of other kinds of applications could be used for different industries such as distribution, transportation, and many others. There must be many ways both telephones and computers could be used together to provide a powerful set of tools for customers to access *their* information. This was my opportunity to talk to people from around the world who were attending the conference, which included end users, and vendors to discuss this technology with them. The results of my discussions were that this would be of interest to many whom I had spoken with.

The interest was there and in a few short months I had been contracted to CallStream Communications Inc. and the adventure began in understanding, marketing and selling this technology. This has been an exciting adventure which continues to this day with many of Scheduled Solutions' clients.

Why write a book on UNIX based Computer Telephony Integration?

The reason to write this book is to provide you the reader with a practical guide to the implementation of these two very powerful technologies. You should be able to pick up this book and look in the tables in Appendix A to identify by industry, and by UNIX platform, the kinds of UNIX based Computer Telephony applications, and be able to see how it is possible for your company or organization to benefit by implementing the technology. As well, you should understand why companies selected a UNIX based product and why it was used in their environment. If you have been able to answer those questions, then our goals to help you have been accomplished. Of course, we value any comments you may have or suggestions for additional case histories.

For those who do not know - what is UNIX?

UNIX is a very mature (over 20 years old) and very modern computer operating system. When it comes to operating systems, they are like a good wine. They improve in quality, and ability with age. Operating systems software controls all the data storage, devices (i.e. printers, modems, and network and user terminal interaction), memory, and execution scheduling of all programs inside the system. Microsoft's DOS, Windows95, Windows NT products and IBM's OS/2 Warp are further examples of operating systems that run on personal computers. Other commercial operating systems are Digital Equipment Corporations VMS for their venerable VAX line of computers, and MVS from the IBM mainframe world.

The UNIX operating system is different from the familiar Microsoft DOS operating system for personal computers, since it was designed to be multi-user and multitasking. This means a single computer can have many simultaneous users, and each of those users could be simultaneously executing more than a single program at a time. For example, a personal computer running Microsoft's Windows95 operating system on a Pentium™ based computer for home use or for the office with 16mb of RAM (main memory), will more than suffice the needs of an individual user to do more than one thing at a time. If the same computer had SCO OpenServer 5 and SCO UnixWare (SCO's trade names of their UNIX operating systems) instead of Microsoft Windows95 it could support up to 30 or more simultaneous users doing more than one task at a time! It is more likely you will find a UNIX system will be performing company wide or departmental tasks such as maintaining a central database application, rather than being used as a dedicated personal computer for word processing tasks. There are as well many workstation systems from Sun Microsystems and Silicon Graphics Inc. and other manufacturers which produces UNIX workstation products, which perform many tasks, but traditionally have been used in academic or scientific research or government systems and the military. These workstation products typically have a lot of computing power to solve highly technical application problems. Moreover, companies like Sun Microsystems, have been successful in moving their product lines into more commercial application businesses with large departmental and enterprise servers that rival large mainframe computers. Sun Microsystems move into this market has brought about new software technologies for this market, which have included their XTL software for computer telephony.

Unlike other operating systems, the internal structure of UNIX is widely known and most if not all off its specifications and software programming interfaces are publicly known. It has been licensed to work on very many different computers from different manufacturers over the years. This has been a benefit as well as a problem. It meant the operating system is well known for its high user capacity and reliability, and that it was used by academia, the military (one military project used UNIX technology to spawn the beginnings of the Internet) and all sorts of commercial applications.

Early in its history, a somewhat limiting problem with UNIX was portability of programs between computers, since each computer manufacturer had to make changes or special extensions to the operating system kernel to accommodate their computer hardware architectures. This in turn forced modifications of software programs if the they had to be moved from one computer manufacturer to another computer of another manufacturer. This process is called porting applications. The term "porting" came from moving cargo from one ship to another, reflecting the arduous task that it entails to do so. It meant some modifications were made to programs for one computer to run on another. This has been finally resolved with a unification of the UNIX operating system specification, which all manufactures must adhere to, in order to use the trade name of UNIX. This process started in the late 1980's and was finalized with industry wide standards accepted in 1993 through to 1995. As well, the special programs called compilers that translate human readable instructions in programming languages such as 'C' to binary computer instructions, now followed these standards to allow maximum program portability between systems. Because of UNIX's open and well understood architecture, it has allowed the operating system to be even more widely adopted for commercial use. UNIX can be used for multiple functions all at the same time, such as being the database server, the networking server for Local Area Networks (**LAN**s) and Wide Area Networks (**WAN**s), direct terminal connectivity, and as we shall see in this book - telephone control and computer telephony integration server.

This is yet another reason UNIX has been widely adopted in commerce. UNIX systems are often used to be the 'glue' within organizations between older and dissimilar types of computer systems. One other advantage of UNIX is that it is not owned by one single dominating vendor. Unlike Microsoft's operating systems, the UNIX operating system is available from a wide range of computer manufacturers, allowing for competitive choice for its purchasers. UNIX now enjoys standardization, portability, scalability, and most importantly freedom of choice of implementation for its customers. UNIX offers something that customers of computer telephony applications most desire, which is high capacity, reliability, stability and scalability.

Because of the difference in software engineering of the UNIX operating system, a similar application written for another operating system such as Microsoft DOS may not be able to handle the same number of simultaneous telephone calls as a UNIX system. In fact DOS was originally engineered for a single computer to perform one task at one time for a single person. Many software programs that run on DOS try to *mimic* multitasking, but inevitably they can and sometimes do fail causing either the program running to shutdown, or worse, stop the computer from working. Mission critical systems such as medical and telecommunications rely on UNIX based systems for their dependability. This is one reason Microsoft has embarked on Windows95 and Windows NT operating system strategies to overcome the limitations of DOS. Computer operating systems can be analogous to workbench tools. You have to use the right tool to efficiently and easily complete the task at hand. UNIX is a powerful tool, that with the appropriate training, is an excellent computer telephony platform.

What is Computer Telephony Integration?

"Computer Telephony represents the cooperative merger between telephony and data communications. It is a form of telecommunications that concentrates on the movement of both encoded and voice-related information from one point to another for the purpose of automating transactions between machines and humans, or between machines"[4]

5

Computer Telephony Integration provides organizations with a new level of customer service, while at the same time providing

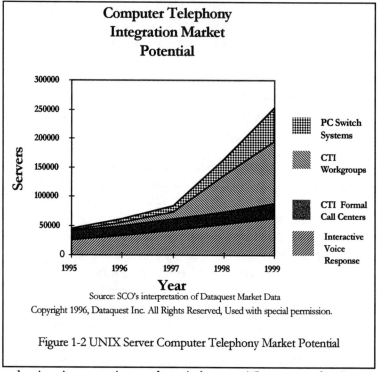

Figure 1-2 UNIX Server Computer Telephony Market Potential

reduction in operating and capital costs. The cost reductions are very measurable in terms of reduced resource expenditures such as telecommunication, computer and manpower costs. Customer service that is provided through these systems is much more flexible and customer directly accessible. This translates to happy customers who will want to purchase from the company again, with the resulting sales and profits. This all adds up to a competitive advantage in the marketplace. For this reason, some of the case histories included in this book have been released with changed company names, since they did not want their competition to know how they are servicing their customers. In fact, some of the case histories show that companies have implemented this technology to obtain this competitive advantage, without a base line of cost justification. Their justification for the systems was just the ability to service the customer better than their competition, rather than trying to provide a cost justification analysis. This demonstrates how mission critical and important these systems are. This is not a technology that is all hype and no substance. "It is a $3.5 Billion US, 30%-a year growth industry".[5] It is also the platform for the much heralded merger of technologies that we now use in every day life. Today we use separate systems such as the telephone network to use voice and fax, we use cable TV to transmit video, and networks such as the Internet or LANs or WANs to move data. As we move closer to the year 2000, we will find that these delivery systems will combine on digital circuits, and computer telephony integration will follow or should we dare say lead, this integration. What we have included here is how our peers have been able to use the technology now and plan to use it in the future.

As shown on Figure 1-2 UNIX Server Computer Telephony Market Potential on page 6, the major growth in the computer telephony integration market is going to be in major new developing areas such as computer telephony workgroups and personal computer switch systems, along with steady growth in the traditional areas of computer telephony such as interactive voice response and call centers. Computer telephony workgroups is an area of importance because of emerging standards in both the computer world, and the telephony world that are allowing organizations to provide the level of customer service that have been in the expensive domain of formal call centers.

As a business person looking to leverage your existing computer and telephony assets in your company, then this figure shows where the potential of growth and implementation is going to be best invested. You can use interactive voice response technology today which can provide a dramatic cost reduction in terms of labor costs, and provide around the clock customer information. Then you may want to investigate and develop a computer telephony workgroup strategy, which can provide a lower cost call center support to provide an even higher level of service. As a reseller of the technology, the same strategy also can be of value. Personal Computer switch systems will help leverage the computer telephony integrated workgroup since the telephone switching may reside within the UNIX system, rather than in a traditional private branch exchange.

The objective of the book is to provide the reader with a practical understanding of how this technology has been used on UNIX based systems. Much has been written about all types of technology that have been invented to aid in the integration of these two titan industries - computing and telephony. Most of this technology includes a litany of "TLAs" - Three Letter Acronyms. (We have provided a Glossary for those we could not avoid.) This problem is especially doubled with both industries converging, which adds to the confusion or prevents some from understanding what this technology can do for your company. One of the goals of this book is to avoid the Three Letter Acronyms whenever possible, and seek to understand how our peers have been able to translate the usage of the technology into happy customers and profits. What was needed is to understand as much as possible how in the real world this has been accomplished.

This book contains is a series of case histories that look in detail at many different kinds of computer and telephone integration using the UNIX operating system. Each of these are a bit of an adventure for those who have generously participated in providing material for this book. Because computer telephony integration is considered a strategic advantage in their industries, the end users of these systems need special acknowledgment for allowing us to discuss their successes. In some cases we have had to give fictitious names for the firms to protect this advantage, or have talked about a company as "a major firm in the financial industry". We thank them very much for their support. I hope that you find that their efforts recounted in this book will provide the insight you need to make you a hero at your company by implementing this technology.

---

[4] 236 Voice Processing Applications Page B-6 , by Edwin Margulies, Flatiron
Publishing, 12 West 21 Street New York, NY 10010, ISBN 0-936648-70-8

[5] <u>What is Computer Telephony? - A White Paper:</u>
http://www.flatironpublishing.com/whitepag.html 02/01/96 9:49:38 PM,
Copyright Flatiron Publishing, 1996.

Chapter 1

Basic Concepts of Computer Telephony

These are the Navigational Instruments you will need for your Adventure.

In order to understand computer telephony applications, you will need to understand some of the basic components and typical application concepts. These include the architecture between the computer and the telephone system, database concepts, speech recognition and reproduction, call centers and "screen pops", predictive dialing and computer telephony integration workgroups. We will then explore the different kinds of applications that have been utilizing this technology. The approach we will take is to introduce the telephony concepts that you need. I am assuming that you understand the basics of computer technology, such as operating systems, application software, hardware and networking. At least you have had experience in your business life working with a computer system to do spreadsheets and word processing. The purpose of this chapter is to provide you with the elements you need to understand the individual case histories in the following chapters, since these have been categorized by the typical applications.

This chapter is split into two functional parts. The first part describes the basic telephony components that are needed in computer telephony. This is the topology of a computer telephony system. The second half discusses the computer portion of the equation. Please pardon some review of the simplest concepts, but all building blocks of knowledge must have a solid foundation.

How does a telephone call work?

This may at first seem to be an irrelevant or simplistic question to answer. For those of us who are lucky enough to live in North America, Western Europe and other parts of the industrialized world, we take the telephone very much for granted. It is so common and expected, we are at a loss if there is not one around. I use the words "most places" to refer to where most of the telephones in the world are - in the developed and near developed countries. In many places around the world one needs to reserve a time days in advance to access a telephone. Although I'm sure that things have changed since I visited India in 1977, I was shocked to learn one had to book an emergency telephone call between New Delhi and Calcutta 8 to 24 hours in advance! If you got a dial tone for a regular call, you should consider yourself fortunate. I remember visiting a lawyer's office about 40km (25 miles) north of New Delhi, and being surprised to see the antiquated 1940's telephone locked up in a wooden box, to ensure no one had access to it. Strangely, it did not seem to matter at all, since I could not obtain a successful connection for the short distance between that town and New Delhi. Even today, in Saudi Arabia, one needs to wait 2 to 3 years to obtain a telephone line, even with the Kingdom's very aggressive infrastructure programs.

Nonetheless, the infrastructure necessary for us to call almost anywhere, at anytime, is quite complex. The principles of a telephone call are relatively simple. For the ease of description, the following describes what transpires while you are speaking and listening on a open telephone line. The telephone handset consists of a speaker for the ear, and a microphone to speak into. How the voice is converted to speech at the other end is not as obvious. The telephone microphone consists of a metal plate or diaphragm. Behind the diaphragm there are carbon granules. Behind the granules is a conductive plate that has 2 electric wires fashioned in a simple electric circuit powered by a battery source. This source of electricity in the public telephone system originates at the telephone company's central switching office. This is why you can still use the telephone, when your home or community is suffering from a general power blackout. The telephone company has battery backups and other means of generating power. The military field telephones have a battery as an electrical source of power. At the other end of the circuit is another telephone handset. The 2 wires extend to the speaker or ear piece. The ear piece is constructed very much like the microphone. When we speak, the voice is converted into vibrations on the metal diaphragm. This motion pulsates against the carbon granules, which in turn modulates or changes the electrical current generated by the attached battery. At the other handset the receiver speaker (ear piece) consists of a couple of magnets and a metal plate. The wire is wrapped around the electromagnets. As the modulated current in the 2 wires flows around the magnets, it changes the strength of their magnetic field. This magnetic field pulls against another metal diaphragm plate, causing it to vibrate. When it vibrates it changes the electrical magnetic energy into sound energy, and you hear the voice of your caller.

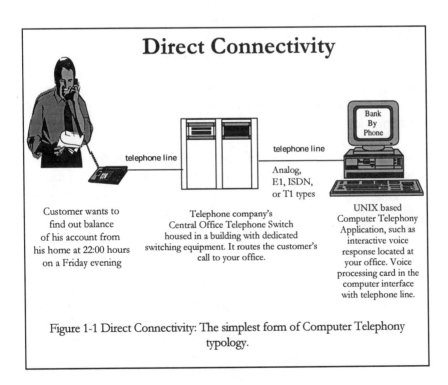

Direct Connectivity

Bank By Phone

telephone line

telephone line

Analog, E1, ISDN, or T1 types

Customer wants to find out balance of his account from his home at 22:00 hours on a Friday evening

Telephone company's Central Office Telephone Switch housed in a building with dedicated switching equipment. It routes the customer's call to your office.

UNIX based Computer Telephony Application, such as interactive voice response located at your office. Voice processing card in the computer interface with telephone line.

Figure 1-1 Direct Connectivity: The simplest form of Computer Telephony typology.

The reason why I discuss this is to show how relatively simple the machine we call the telephone really is - and it has not changed very much in the last 100 or more years it has been used.

Direct Connectivity

Direct Connectivity refers to how the telephone line is terminated in the computer system. A telephone call passes through the telephone company switching equipment in a central office in one part of the city, and connects with another central office switch in the same city or other locale. The telephone line and call connects directly to the back of the computer running a computer telephony application. Inside the computer there are voice processing cards. These cards understand the hardware interface and signaling the telephone line is using.

Direct connectivity is the simplest form of connection you can have, and sometimes most desirable when you are providing an application where you may not need to transfer the call after the caller has used the application. A typical application of this configuration is interactive voice response, such as banking by phone. Sometimes this is an excellent entry level computer telephony application since it requires the simplest form of integration.

The corollary to this kind of connectivity would be for the call to terminate on your own private branch exchange or private switch in your office. Then the calls would be directed according to your switch programming, and eventually would be connected to your application on the UNIX computer system.

Call Flow

Call flow refers to the path along which a call originates and eventually terminates. For example, let's say you want to check the balance of your chequing account. Instead of calling your local branch during business hours, you call a number and the bank-by-phone system asks you for a customer number. (The bank branch had previously sent you information on the service including your customer number and a password.) You reply by entering the appropriate digits using your keypad on the telephone. The system asks for your password and you supply it. The system provides you with a menu of choices. You select option "1" and the system provides you with the balance of your account. Other options could include transaction history, current exchange rates, today's interest rates, balances on loans. It has the ability to pay bills and to transfer funds between accounts, fax you a customized mortgage schedule, or retirement plan information. All this happens with real time data and can be done anywhere in the world, 24 hours a day, 7 days a week. This describes the call flow which on the surface is relatively easy to understand. The call is generated by you and routed through the telephone network to the bank's data processing center. In the simplest configuration these outside telephone lines connect to an amphenol™ block. This is a simple device that provides multiple telephone jack (RJ11 standard) inputs on one side, and merges them so that a wide ribbon cable can be attached to the back of the amphenol. These amphenol RJ11 blocks are nicknamed "harmonica" blocks, since they resemble wide harmonicas. The wide ribbon cable that attaches to the back of the harmonica block connects to a voice processing card, which resides inside an Intel 486 or Pentium™ PC computer running an operating system like SCO UNIX, or Solaris x86 or other UNIX systems. This system has voice processing application software. This controls the voice card playing pre-recorded speech and collects the input from the telephone's keypad. It calls upon database software to find your chequing account information residing on the bank's larger computers, and then speaks or faxes back, or even e-mails the results back to you. In this case the telephone acts like a data terminal, from which a bank employee would have gathered the information to answer your questions. The fax machine is the equivalent of a printer on which the employee would have printed the results.

The Mechanics of a Call

When you pick up the hand set of the telephone you hear a dial tone. What really happens is that you pick up the phone and a telephone company's central office telephone switch detects a change in voltage, and then presents you with a dial tone. The telephone keypad or rotary dialing method are means of routing a call through the telephone network. The central office telephone switch, or (central branch exchange), collects your routing information in the form of pulses or touch tones, and routes the call to another central office switch, which in turn sends a ring voltage to your telephone to make the telephone ring. You pick up the phone, and the electrical circuit between the two telephones is completed, and you can speak and hear each other. For our purposes you need to understand that the voice we hear on the telephone is really an electrical current which is *analogous* to the real voice that is speaking, if you are talking over an analog line as described above. When you talk on a telephone the local loop of wire from your home is most likely an analog circuit. Once it has reached the telephone company's central office switching system, the transmission of the conversation between central offices is most likely a digital transmission. This is where the voice is converted to a series of binary "ones and zeros". This is then converted back into an analog circuit at the other end. This is most likely the kind of transmission in homes in Canada and the United States. Larger companies with many lines will elect to rent digital lines that directly connect from the business to the central office. Some popular kinds of digital circuits are ISDN, T1 or E1 lines. (These are discussed later in the chapter). In some countries in Europe, such as Germany, ISDN services are very common for both business and home.

When you use a touch-tone phone you generate a specific frequency of sounds, which are outside normal human speech. This is important for most applications. The tones you hear when you press the keys 0 (zero) through 9 and * and # are not a single frequency of sound. In fact they are two sounds being generated together. The technical term for these tones is Dual Tone Modulated Frequency or (DTMF) for short. It is these tones upon which most computer telephony applications are built. As well, there is special added information that is being sent along with the call, which are Automatic Number Identification (ANI) or Caller-id, and Dialed Number Information Services (DNIS). This is digital information that is sent within a split second after the first ring on the receiving phone line. This information tells who is calling, and which number they called. If set up properly, these can be valuable tools in applications.

Caller-id

Let me explain why automatic number identification (ANI-hereafter called caller-id), and DNIS (hereafter dialed-number) can be very useful to an interactive voice response system. Let us say a university has a student course registration interactive voice response system. The registered student already received course outlines and guidelines with which they must register for electives, as well as required courses. She calls the system and is prompted for her student id number (i.e. 97813731) and the assigned password. Thereafter, the student would be guided by voice prompts to register for the courses and classroom times, according to the business rules encoded in the logic of the interactive voice response application. In the case of advanced database technologies, these rules are encoded as part of the underlying database. As the student replies to the prompts, the telephone key presses are translated into data numbers or characters by the voice processing card and interactive voice response software, which then can be used to retrieve or write information back to the student registration database.

If caller-id data can be detected and retrieved when the student calls, then time and money can be saved by the student and the

> **Jurisdictional Availability of Caller-ID**
> As an aside, caller-id service is not available in all jurisdictions in the United States, or it is restricted to specific kinds of telephone services. For example, before June 1, 1996, the State of California only allowed caller-id on 1-800 toll free services to aid call centers. Concerns from special groups such as special women groups felt that if a phone call made to the person (i.e. an abusive husband) persecuting them, then this detailed information could be trapped by a caller-id display device, available from telephone company retail outlets or electronic stores. If the woman was in a 'safe house' then the location would be unavoidably disclosed.

university. The reason for this is, caller-id consists of a phone number and possibly other information which is captured after the first ring on the inbound telephone line. Depending on the phone company which originated the call, and depending on the switching equipment they used to route the call, other information can be noted, such as the date and time of the call, and even first and last names and address of the owner of the phone from which the call is originating. The full name and address information is not available in locales. For example a string such as "416-555-0000 1996/04/15 15:43:03 Joanna Dobson" may be captured by the software. This number is checked against the student registration database, and if it matches, a couple of things can happen within the limitations of the business rules. First of all, if there are no duplicate phone numbers in the database (i.e. student roommates sharing a residence), the system could load any business rules from the database, eliminate a prompt asking for a student number, ask for the password only, and then follow the business logic for the remaining of the registration for this student. On the other hand, under the same circumstances, if the university has not received tuition payments, the call can be transferred to the business office. A computer screen of information for this student could be presented to an university staff member and then have the student's call transferred to that staff member. Alternatively, the staff member would receive the call with a pre-recorded message indicating the nature of the call, before the student's call is transferred to them.

The amount of time saved for one interactive voice response prompt may be anywhere from 5 to 10 seconds, depending on the amount of data being keyed by the student on the phone. You may say 5 to 10 seconds is not very much, and why bother! Well, multiplied by a student body of 20,000 students, this calculates to about 55 hours of telephone time within a period of 1 month. If the students called long distance, or the university provided toll free calling to the system, this would translate to about $850 at a rate of 25 cents a minute for the 800 or 888 toll free service. As well, the screening of the callers (i.e. those that have not paid, or who are no longer eligible to enroll), provides better financial security.

Overall system security is maintained if the interactive voice response software can keep a log of the callers and their frequency of the calls. For example, if a mischievous or disgruntled student wanted to tie up the phone lines to the registration system at a peak time, using his personal computer modems to dial randomly and constantly to flood calls into the system, then the interactive voice response software could immediately report that this is happening. The system administrator could then tell the interactive voice response software to not answer, or immediately drop all calls from those phone lines. The system administrator can look up the phone number in the university database and then take the necessary action to prevent this from happening.

Although not an obvious saving to the university, the 55 hours of telephone time saved translates into a reduction in the amount of voice board hardware and software that the system needs. If the caller-id saved this amount of time at peak registration, then it could save the university from needing 2 to 4 telephone channels. This represents an additional saving of about $2,500 to $4,000 in hardware and software costs.

Dialed-Number

The university could offer another interactive voice response service, such as a library line. This could be used to search for availability, or extend the loan period for books or reference material. The caller could enter the library reference number or ISBN number to identify the book or material. Normally, additional voice board hardware and software, and on-going telephone line costs would have to be purchased to accommodate the new application, since these added telephone channels would be dedicated to the application. If the interactive voice response software could utilize a DNIS (dialed-number) function, then these additional costs could be eliminated.

When the port on the voice processing card answers ("goes off-hook") after the first ring, it collects the dialed number information, and based upon the identifying number, dynamically load the appropriate application.

Hunt Groups

The inbound calls are not stacked in a queue behind each other unless there is a special mechanism within the switch, called automatic call distribution, that puts calls on hold and prioritizes them. Automatic call distribution is discussed below.

A hunt group is a set of telephone lines that are organized in a logical group or subset of lines. The group can be as small as 1 line or extend in groups of many lines. Remember, a telephone call you originate travels through the public telephone network. The major role of that network is to route calls, via the telephone company's central office switch, to the university. In an organization of the size of a university, there is usually one or more smaller private branch exchanges, which routes the telephone call traffic throughout the university. For argument sake, lets assume the university has 24 telephone lines assigned to the interactive voice response system. These lines originate directly from the telephone company's nearest central office private branch exchange and terminate into the back of the university's own private branch exchange.

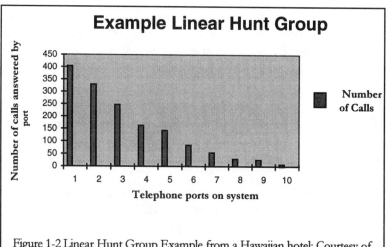

Figure 1-2 Linear Hunt Group Example from a Hawaiian hotel: Courtesy of CallStream Communications Inc. Used with permission.

For economical and technical reasons, 24 lines of telephone is best served (in North America) with a special type of telephone cable called a T1 line. This is a twisted pair cable that carries 24 digital telephone lines. In this case the fact that they are digital versus analog does not really matter. The most important thing that digital line services usually have is caller-id and dialed-numbers. This type of telephone cable can terminate directly into one of the university's private branch exchange, or in some cases directly into a special T1 card that resides within the computer that holds the voice processing cards. When the calls arrive from the outside world, they are sent to the hunt group in the university's private branch exchange or T1 card within the interactive voice response computer. A specific subset of these 24 channels can be dedicated to a specific function. Those lines are designated "a hunt group".

Hunt groups can be designed to be either linear or circular. A linear hunt group means that when inbound calls are received, the private branch exchange or T1 card will then start looking from the first telephone port to the 24th telephone port on the voice processing cards, in a sequential fashion, hunting for an unused telephone line to answer the next call.

Using a linear hunt group the first few ports will always answer more calls then the last ones. The side effects of this type of hunt

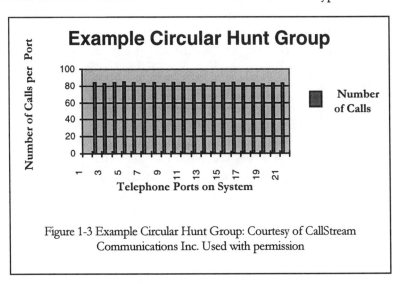

Figure 1-3 Example Circular Hunt Group: Courtesy of CallStream Communications Inc. Used with permission

group is that it usually takes more time to find the next available port to answer the call, when the system is very busy. On a busy system the caller will hear more ring tones before the call is answered. Usually the reasons for a linear hunt group implementation is that this is a restriction on the manufacturer of the university private branch exchange, and to allow any telephone call to get a busy signal. The system administrator may have such a large call volume, he may want to ensure that the call will receive a busy signal once all the telephone channels have been tried. If all the ports are busy, then subsequent calls will also have received a busy signal. A circular hunt group will search sequentially from the first to the Nth telephone port of the system looking for an open telephone port to use. If unsuccessful after the first pass on all the lines the call goes unanswered, it will start the process over again, and again, until the call is answered by a free port. The number of cycles through the hunt group is determined by programming the private branch exchange, or by setting software parameters. If the call can not be answered after the predetermined number of cycles through the hunt group, then the caller will receive a busy signal.

When one first looks at a computer screen showing the activity of ports on a large system, it is fascinating to watch the hunt group in action. The status screen will show that there are many different calls happening at once, and each is in a different location within the application. Thus they are in various states towards call completion. If you are watching a busy system you will suddenly see a telephone channel go "Next Call" (status message is totally dependent upon the software in use) indicating it is waiting for another call. Then in what appears a random act, the unused ports go into use. Related to hunt groups are the port utilization management reports that show you the number of calls per hour by port, and the number of call minutes per port. These are key to show when the system is the busiest during what periods of the day, while the second shows the utilization time during the day. These reports are important to indicate if you require more hardware and software to cover the increased load on the system, or to make a decision to add another application to an under utilized system.

Figure 1-4 on page 23 shows two reports that have been graphed for readability from a resort hotel in Hawaii, USA. It shows a system that is very dormant at night, and continues to increase in call volume as the day progresses to just past noon, dips a bit, climaxes by 5 to 6 PM, and then dramatically drops off. This shows the peak usage of the system, which gives you a true view. If you just viewed the linear hunt group for the same time period and location, you would never know how the guests at this hotel actually are using the voice messaging system. With a linear hunt group you would see that your last ports are not peaking as high as your first ports, giving you an idea that calls are being answered, although it may take more rings than usual.

With a circular hunt group, which evenly loads all the calls on all the telephone ports in the hunt group. There is no self evident indication or hint of underlying trouble in the utilization of the system. So the Calls by Hour and Calls Minutes by Hour reports really show the depth of how busy the system has been.

As you may suspect, the call minutes per hour report reflects the same graph as the calls by hour graph, but with more refined detail, as to how many total minutes in a single hour is used on the

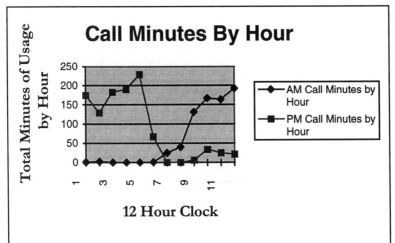

Figure 1-4 Call Minutes by Hour: Hawaiian hotel. Courtesy of CallStream Communications Inc. Used with permission

system. Since there is more than one port and since this report is an accumulative total of port usage, then it is possible to have more than 60 minutes of utilization within 1 hour. Hunt groups and hunt group utilization statistics, provide you with the ability to route calls to a specific set of telephone ports to your computer telephony application, and shows how well your system is tuned and able to be take on more load.

Automatic Call Distribution (ACD)

This is a function of the private branch exchange or of a central office switch. It holds the inbound callers in queue until the next available person is able to answer the call. This is the typical way to keep callers on hold whenever you make a reservation with an airline. It is an important concept to understand, since it is a key element in computer telephony integration of call centers and computer telephony integrated workgroups. The main objective is to reduce the wait time any customer may experience, since it costs the company money, (especially if it is a toll free call), and it wastes the time of the customer. Computer telephony integration applications strive to reduce that wait time as much as possible, without having to add extra manpower in the call center or workgroup. We will discuss the details of this below.

Private Branch Exchange (PBX) Integration

Private Branch Exchange integration refers to the ability to integrate a computer system running a computer telephony application to signal to a private branch exchange. This is a very important function. Not understanding this could really make your life difficult. What in the world is a private branch exchange? It is the piece of *proprietary* telephone equipment. It is the special box in the telephone closet that allows you to make telephone calls within and outside your office. It is usually from vendors such as AT&T, Comdial, Ericsson, Mitel, NEC, Nortel, Rolm and other manufacturers. Your office telephone connects to the ports on this device in the wiring closet. When you pick up the phone and have to dial an access number such as "9 for an outside line", you are informing the private branch exchange in your office to attempt to find the next available telephone company's line. The telephone company's central office exchange (or "switch") provides you with dial tone, so you can dial the number you want, and accordingly routes your call to the office next door, or if necessary around the world.

The private branch exchange equipment is sold with a ratio of inbound telephone company lines, to the number of extensions hanging off the system. So if a telephone specialist or "interconnect" who sells telephone systems says that the private branch exchange or "switch" you are looking at purchasing is "12 by 36" system, this means that the equipment can support up to 12 incoming telephone lines from your telephone company, and provide you up to 36 telephone extensions inside the office. This means that only 12 of the 36 telephone extensions can be used to make or receive an outside call.

➤See *Appendix B: System Sizing Tables, on page 333* to help size your computer telephony application.

This is obvious, because you only have 12 telephone company lines to make or take calls from the outside world. As you have noticed, the number of available telephone lines and the number of potential extensions is based on statistical calculations. As with an office system, the telephone company's switching equipment depends upon this same statistical ratio. So if everyone in a residential and office area picked up their phones at the same time, only a fraction of the users would actually obtain a dial tone. Unlike a telephone conversation that lasts a few minutes, a computer modem connection to the Internet or another computer can last hours at a time.

This has been seen as a potential problem by the large telephone carriers. An article in *The Economist* discussed the issues that the "Baby Bells" are trying to paint, in order to have regulatory agencies in the United States put some restraint on Internet based access, and especially Internet based telephony, by using this scenario: "WINTER 1997: on a snowy afternoon in New York city, an elderly grandmother feels sick and reaches for the telephone to call an ambulance, but there is no dial tone. She had always paid her telephone bills on time, just to ensure that help was at hand. Now the telephone system has let her down - all because a few dozen Internet junkies on her street are hogging all the available telephone circuits."[8] Although this is a futuristic situation, it shows the influence different kinds of traffic can have on the switching equipment at the telephone company's central office.

If you are an UNIX practitioner, the word *"proprietary"* may not be in your linguistic vocabulary, or even a word you want to mention in polite company at a UNIX trade show. UNIX based technologies are standards based. Many years of effort have been put into this part of the computer industry so as to allow many vendors and product developers reach the widest possible audience and market. The tenants of interoperability, scalability, standardization of operating system software and application interfaces to the operating system, and hardware are held as truths today, by which we all on this planet are benefiting. The Internet is built upon those very foundations. Even manufacturers such as Microsoft, Novell, IBM and a host of others, have taken from, or had to adopt these standards for not only the Internet, but for many other types of system software, hardware and applications. If you 'surf' the Internet using Netscape you should be singing the praises of standards based TCP/ip networking that has it origins in the UNIX industry. You may enjoy the "new" benefits of Windows95, and Windows NT allowing Microsoft applications to do many things simultaneously, and not crash your system. If you are so inclined, you may want to get down on your knees and thank the early UNIX operating system developers who developed "preemptive multitasking" and "threads" that allow you to do this. Well, I better end this portion of my sermon before the UNIX choir starts to sing.

I'm sorry to say for the UNIX practitioner, in the world of computer telephony, the word "proprietary" has to enter your vocabulary - and yes - it is even polite to say it. The private branch exchange manufacturers have agreed on the basics of making and placing a call, and other signaling technologies, such as Integrated Switch Digital Networking (ISDN), and T1 and E1. As for any voice or data that is rattling inside the private branch exchange, or at the telephone company's central office exchange, how that is switched, transferred, and handled is completely up to the designers of the switch manufacturers.

At the 1995 SCO Forum, the yearly UNIX trade show for the Santa Cruz Operation (SCO), Harry Newton, the publisher of this book, gave an interesting speech of how switch manufacturers build their systems. To fix a simple annoying function on his telephone set that Harry had complained about, he was informed that it would take 1 month to correct the problem, and 1 year of regression testing to test the change in the switch's telephone software.[9] The manufacturer had not used an operating system inside this device. The software was one large program, which had to be completely tested from keel to mizzen mast. If they only had used an operating system such as UNIX the change would have been a snap! They would have only needed to test the software piece that was changed, and not the whole system. Harry further went on to say, that even different products within switch manufacturers have completely different software from the other.

Hence the UNIX tenants of scalability, interoperability, and openness in standards with product lines usually do not apply. (My fellow UNIX practitioners are cringing by now.) The expensive telephones used with the small office switch that give you 12 telephone lines, and 36 extensions in a small branch office, may be incompatible with the same manufacturer's larger switch in the head office. As a business manager, this is why you own a $250 telephone you only use to "press 9 to get outside line". Maybe you've been become a bona fide 'telephone power-user', since you can perform "conferencing" and "transferring" of calls. You have achieved this great feat, since you have taken the time to learn all the arcane codes on your telephone set to make these functions work. If you move to a different company which has a different switch manufacturer for a telephone system, you will have to learn a new set of codes all over again - how nice. Making a call is very simple, and the technology needed is not very different from what Alexander Graham Bell used 100 years ago. What makes the telephone on your desk cost $250 is the proprietary signaling that the switch manufacturer has introduced in order to signal the telephone to conference calls, transfer calls, and be able to set message waiting lights. These phones are sometimes called 'digital' or 'intelligent' phones.

The only difference between the phone that you have at your home and the one in your offices is, the office phone has a small computer chip that aids in the signaling of its status back to the switch. This type of chip is called a Codec chip. (As a business person, I promise you will never need to remember the name of this chip.) Each switch manufacturer has a proprietary set of instructions written in that chip. So this makes it impossible to take an AT&T telephone set, and use it with an Nortel Meridian switch, or other phone switch. The only common standard for all telephone switch manufactures is essentially the same technology that Alexander Graham Bell devised so long ago, and it is called "Plain Old Telephone Set" - (POTS). Because of this incompatibility, there is a burgeoning part of the computer telephony industry that is addressing this integration area. It is called computer telephony integrated workgroups. This integration of desktop computers with the telephone switch uses - wait for it - standardized software interfaces.

I mention the switch's set of codes on purpose, since this is truly the crux of the problem, when it comes to integrating a computer based telephone application to a switch. I use the word 'crux' on purpose since it has its Latin origins in the word 'cross' which is related to the word 'crucified'. You have really three types of technology vendors when purchasing UNIX based computer telephony applications, such as interactive voice response, voice messaging, call centers and others:

➤Want to learn more about becoming a telephone power user using your desktop computer? See Computer Telephony Integration Workgroups on page 72.

    1. You can purchase these applications directly from the manufacturer of the switch. This has always been the traditional source of products. Computer telephony applications from such vendors are usually a "soup to nuts" business approach. You purchase all your computer telephony needs from one vendor such as a AT&T. You are pretty well guaranteed to have excellent integration between your computer telephony applications with the switch and its ability to transfer calls to the appropriate destination. Like the mainframe computer manufacturers of days of yore, they are able to quickly and efficiently service nearly all your needs. This suitor may, or may not, have all the answers from the UNIX based computer application side of the equation, and may not be as familiar as you may like with that particular environment. The downside to this arrangement has been, in the past, that you may be locked in with that vendor, and that it is not easy to integrate other products into your solutions set. This may take away some of the freedom of choice you may want to respond to your ongoing business requirements. Being wed to a single vendor can be very high in the long term.

2. You can purchase products from 3rd party software and hardware vendors to integrate with your switch. This book's case histories describe the successes these companies have had in providing the solutions, using many different interfacings with many different switch manufacturers. They can provide you with very competitive products which can integrate directly with your existing UNIX based applications. The applications cover the full spectrum of computer based applications. This type of vendor has moved the computer running the voice mail, interactive voice response applications and other such applications outside of the telephone wiring closet, where the switch usually resides. The vendor has either placed a dedicated server talking to your existing UNIX based applications, or they may even be hosted on the same UNIX computer running those applications. The key thing about such vendors is that they have taken the pain and trouble to work out what the switch manufacturer's switching codes are, and how they effect the behavior between the switch and the computer telephony application. They have achieved this either by trial and error, or have become an authorized developer of software for a specific switch manufacturer, or both.

The switch manufacturers are recruiting these vendors with specialized developer programs to enable more 3rd party applications to run smoothly with their switching equipment. Like the mainframe computer manufacturers in the late 1980's and early 1990's, the switch manufacturers have seen the writing on the proverbial proprietary wall, and realize they can no longer own the whole traditional telephony business model. Customers are demanding an opportunity for choice they have seen and experienced in the computer revolution. By adding 3rd party applications the switch manufacturers are able to sustain their business model further by adding value the customer wants to buy. Moreover, these switch manufacturers have even adopted 3rd party developer's software as their own and actively resell it to their own client base.

In the end, the company who purchases these systems and is able to integrate them into their current computer systems with transfer of the key and critical knowledge, will allow them to be more flexible, and have more freedom of choice. The encouraging thing about these vendors is, they have worked with many different switch manufacturers over the years. Hence, if your business outgrows your current telephone switching capabilities, and you need to change manufacturers, you will most likely be able to retain your UNIX based computer telephony applications and all the ancillary and expensive human investment that goes with it.

Even if your UNIX computer telephony application vendor is not sure about integration with a particular switch, you may elect to have them do an integration attempt on a conditional, trial and error prototype system. It may be worth it just to get the freedom of choice you may want to have. This may involve the software developer technical representative coming to your office and hooking up a portable system to your switch, to see how it does, or does not communicate. The need to come onsite is a valuable one, since that person will become familiar with the operations of the office, and more importantly, be able to work with the switch as it is currently configured by your telephone interconnect. You should have your interconnect or your in house telephone communications technical administrator be present during the trial, in order to have the switch to computer telephony application interface programmed between the systems. For a voice messaging system this may take about 1 to 2 days of cooperative effort, to get these things to run smoothly enough to allow a limited number of users to try out the system for a suitable length of time. Once the basic interface is up and running, the users will experience problems, and they should be technical enough to be able to document them. Items such as having just received a call, and then needed to transfer the caller into someone else's voice mail box, and have the transfer not successfully completed, or the message waiting light not turn on, are some of the issues that you may discover with such a prototype. There will be different symptoms for different functions depending on the kind of application being installed.

The key to success is the cooperation between the switch manufacturer's interconnect and your UNIX computer telephony software vendor. In one case history, which will remain nameless, a technical developer who purchased a UNIX computer telephony development system, said it was a very frustrating experience for him until he was able to put both the UNIX computer telephony application developer technical representative, along with the technical representative of the switch manufacturer, and himself into the same room for a day - with the doors locked. Calls were not going where they should have been going, and the two vendors were saying the problems were each others fault. This was resolved when they were able to trace the problems together in a cooperative fashion - face to face working with the equipment together. The technical developer was able to get his programming of the call queues much easier after that day. In hind sight, he said he should have made sure the integration was much tighter between the two vendors, and that the computer telephony vendor had at least a couple of sites where his type of application had been already installed.

Moreover, he feels the company really benefited. If there is a sudden new need to be addressed in their customer service organization, he said he could now develop a new computer telephony application, with full integration to his corporate database, within about 2 weeks. He further added this was a capability his company could never have if they had decided to purchase a single vendor solution. This was the critical reason why they went through this process. Instead of trying to find a vendor with a list of reference sites, they may have tried the prototype trial and error system, before major development resources had been committed. His tale of woe and inevitable success highlights another important issue.

The joint cooperation of your computer staff with the telephone communications staff in your organization, or in a smaller business, the telephone interconnect is essential for successful project completion and delivery.[11] I've personally been selling interactive voice response and voice messaging on

UNIX for almost 4 years out of my 8 years selling, marketing, and even some light programming of UNIX systems. As a manager of your business, which has both professionals on staff, you may be surprised to find these two groups do not interact very much, and really do not understand very much about each others business. It is important to have them cross train as much as possible, so that their viewpoints and biases are adjusted to the overall goals of the organization, which is better customer service. When you have to use your telephone interconnect as part of your team, you may find the response interesting. On more than one occasion, I've seen the interconnect sales and management become less than cooperative, and in one instance just plain obstinate to his customer. This may be expected, because the interconnect may not make a lot of money on the actual telephone system, since it is a near commodity item. They may make more money when he installs other enhanced products such as voice mail and interactive voice response, and call center services such as automatic call distribution software on the switch. The interconnect usually represents the more traditional vendor as described above. He may see the UNIX based computer telephony software vendor as competition, and taking bread from his table. This is an understandable reaction from his point of view. In one case in North Carolina, an interconnect told his customer that he was going to charge an exorbitant rate in the thousands of dollars per day, if he had to consult, and help reprogram the switch for the UNIX software vendor's installation. The customer called the switch manufacturer, and asked for a name of some different interconnects. The customer wanted choice, and if he had to, he would either find another person to service his switch and work with his UNIX software vendor, or find another switch manufacturer who would be more open to such integration.

The return on investment these companies have obtained has been in the millions of dollars per year! After reading the paragraphs above you may be scared off, and put this book aside in the to be read pile in the corner of your office. Don't! The effort is well worth it, and the guidelines I discussed above will help you maintain project control and meet your goals, and you will not be crucified. Although there are no direct standards for some of the interfaces, these companies will give your firm the flexibility and preservation of investment you need to deliver excellent customer service.

3.   Our third vendor is the one who provides an interpretive middle software layer, which is called "middleware" in the software industry.   There are a handful of vendors such as Dialogic and their CT-Connect, Genesys Labs and their T-Server product, Aurora, Nabnasset and others who have taken the trouble to analyze all the different major private branch exchange equipment and all their proprietary codes. They write software that will allow 3rd UNIX computer telephony vendors (our second category of vendor), to transfer the call with requests like the following:

> "Transfer this call with this screen of information from the Sybase database on the UNIX host computer, when the next available customer service representative is available. If there is no one available in 10 minutes please transfer the call to this other high priority queue in another city. If the caller wants to leave a message I'll take the call back and transfer him to voice mail. After the message has been completed, I'll ask you to turn on the message waiting light."

4.   In this request dialogue, the computer telephony application does not even care what switch is being used, nor does it have to know anything about the codes used to make all this happen. The middleware accepts the commands, does all this nasty work interfacing with the switch, and then reports back to the calling computer telephony application with a simple set of software commands. Software like Genesys, is a product that you as an end user may need, but may not be able to directly apply in your own environment.   Usually, the product is integrated by the computer telephony software vendor, and is tested with a solution. One company that offers this kind of solution is Scopus. They offer "off the shelf" help

desk, sales management, and other applications that integrate directly with Genesys, and Aurora. I mention Scopus, Genesys, and Aurora only because I have some limited personal knowledge about their products, and by no means the intent is to endorse their product or companies The purpose is to show that the growth of software products from companies such as Genesys, are providing a much needed interface, which provides you with a greater degree of choice and integration.

Fortunately, more and more standards are evolving to make sure that the capability and integration between the private branch exchange, and the computer telephony application system are making the use of this technology much easier. Yet, the allegory of the "chain is as strong as its weakest link" applies to this type of integration. If your ability to communicate between the switch and your host computer is limited, so will your application capabilities.

What does the future hold for private branch exchanges and their manufacturers? The advent of the personal computer (PC) switch, is the next logical sequence of events. There has been limited success of such products on the Microsoft DOS, and MS-Windows 3.11 operating environments. Some have come and gone, while others remain. These have not been a threat to the main switch manufacturers, in that they try to solve a solution in the small business market, with a single card supporting 18 or more telephone lines. I would also hazard a guess that the products in the past have been limited by the marketing issues, such as prospective customers would normally expect to purchase telephony switching equipment from their interconnect, rather than from their computer supplier. On the horizon, you may see new PC switch products that will be more readily accepted in the market, since the computer distribution channel is becoming much more computer telephony savvy, and be able to handle the switching requirements and bundle it with their knowledge of databases, client-server and network computing. These products are available in the UNIX market today, and we examine some of the solutions in the chapter on PC Switch.

➤ See Case History: The "Intranet Long Distance" - AlphaNet's Mondial Network on page 241 as an example of PC Switch technology used in international long distance.

Then there is the Internet, which has a wonderful habit of changing all the rules. The merger between computer telephony and the Internet has arrived, and it completely bypasses all the standard switch based telephony, which provides the best of both merged worlds. This is one of the most exciting areas in the computer communications industry. Instead of connecting your voice and fax communications from the telephone company, you are able to provide connectivity directly from the Internet into your UNIX computer system where all the data is stored. These are very interesting times indeed.

Types of Switch Signaling

Switch signaling is instructions to handle a call, and how to treat it as commanded by the computer telephony application and the switch. Although, as previously mentioned, there are emerging standard protocols, the following are the old standbys which achieve the desired effect. Most of these really describe how the product interfaces and does the signaling, rather than the actual signaling itself.

INBAND DUAL TONE MODULATED FREQUENCY (DTMF) SIGNALING

➤ *HINT*

In response to an autoattendant for a mailbox number may not be the same set of digits representing the physical phone extension attached to the switch.

Inband DTMF is a very common form of switch signaling, and is sometimes the signaling method of last resort, between your computer telephony application such as voice mail, and, or interactive voice response system, with a private switch.When you phone a company that has an autoattendant to answer all the inbound calls, have you ever wondered how the call is handled? The call arrives at the private branch exchange, and the call is then forwarded to a hunt group of lines that terminate on the voice mail system running on a UNIX system. The call is answered by this system, (note - not the switch), and voice mail software looks to see if there are any special instructions encoded on the line as special DTMF signals. If not, it then plays the autoattendant greeting. The voice mail system requests you to enter an extension number, which you do using the touch-tone keypad on the phone you are calling from. The voice mail software then collects those digits, examines a database for a corresponding switch extension number, then looks to see if there is a free line between the computer system and the switch. If there is a free line, the voice mail system then picks up the line (goes "off-hook"), and gets a dial tone from the switch. It then sends a special string of touch-tones represented by the numbers on the telephone key pad such as *#7546. The "*#7546" may mean that the switch is to transfer the call to 7546". The switch now dials that extension. After trying 3 or 4 times to reach that extension without response, the call is transferred back to the voice mail system hunt group with "*7546*". The voice mail system answers this internal call, looks for the special codes, and listens to the tones "*7546*" to indicate that the party at extension 7546 is not there, and please play the mailbox owner's personal greeting, and then accept a recorded message.

The signaling here is completely done by the tones being played back and forth between the equipment. It is not very advanced technology, since the signaling is no different than one beating on a drum to provide information. The physical interface is just a regular analog telephone line. There is no other way to say what to do with the call.

In the description above we demonstrated something that should be explained further, which is the difference between a "supervised transfer" and an "unsupervised transfer". These kinds of transfers are totally dependent upon the ability of switch's software and to some degree on the limitation of the computer telephony software.

You may think at first that the term "supervised transfer" would be the best way to handle a transfer, while an "unsupervised transfer" would mean that there would be more overhead. It is quite the opposite. If you were purchasing a voice mail system, (or other computer telephony application that required calls to be transferred somewhere else), then you would want to try to get unsupervised transfer capability. What we described above was an unsupervised transfer. Once the caller had entered the extension number he or she wanted to reach, the system went off hook, and sent the "*#7546" to the switch, *and then hung up the line*. The switch then did all the dialing of the extension without the voice mail system's direct control, and the caller heard it ring the phone 4 times with a no answer result code of "*7456*", and then ring the voice mail system to tell what happened. In this case the voice mail software did not have to stay on the line like a receptionist to listen to the rings to find out if the call was successfully answered.

Why as a business manager would I care about kinds of transfers? It boils down to minimizing your costs, and getting the most out of your equipment. Do not forget your UNIX based computer telephony application is best suited to handle many things at the same time. With unsupervised transfers, the telephone lines and subsequent channels are being used more efficiently, and therefore not being tied up monitoring the success of calls being transferred. Hence, your voice communications card in the back of your system will handle more telephone traffic. This cuts down on the cost of software licenses, and hardware you need to purchase.

OUT OF BAND SIGNALING

Out of Band Signaling takes all the messaging between the switch and the UNIX system running the computer telephony application and places it on a wire between the two systems, rather than within the message transfers themselves. Out of band signaling usually is much faster since none of the time the caller is on the line is being stolen by transmission of touch-tones as in inband signaling. Generally, the signaling is provided using a software interface, which drives a common standard piece of communications cable directly connected between the computer and the switch.

An example of out of band signaling is NEC's MCI™ interface for their NEC 2400, (enterprise wide private branch exchange), and 2000, (small to medium size system,) and others in their product line. The MCI interface is a software interface that is unique to NEC's equipment. It is similar to how other manufacturers implement such interfaces. The MCI interface was ported to UNIX by CallStream Communications Inc. of Oakville, Ontario, Canada for NEC. The interface allows for complete quick call signaling via standard serial communications (i.e. modem-like) link between the systems. Transfers of calls are unsupervised, and therefore faster and easier to maintain. Full integration means minimized problems with such items as message waiting light control, and out bound

38

dialing features of an application. Compared to regular inband touch tone interfaces the speed of transfer is much faster. Many switch manufacturers have placed these out of band signaling schemes as published specifications for computer telephony developers, in order to add value. Of note in the NEC 2000 IVS product, there is even a TCP/ip connection capability. This is one manufacturer who realizes the need to allow communications with its equipment as part of the wider office LAN, and WAN of the corporation.

SMDI SIGNALING

SMDI signaling is another form of out of band signaling which deserves its own honorable mention. It is a Bellcore standard, which indicates how packets of information are formatted and transmitted between devices. This standard has been adopted in some limited fashion within the industry. Of note, is a company in Buffalo, New York, called VoiceBridge Technologies that has used this signaling method to help computer telephony developers obtain a common interface for communication with various switch manufacturers. They sell a product called VoiceBridge, which talks to the switch in its native proprietary set emulation, and then translates that as output into SMDI packets, which are transmitted via a serial (i.e. modem like) cable.

When a call arrives at the switch, the call's proprietary signaling for that switch is transmitted to the VoiceBridge via telephone extensions, and is converted internally to a SMDI packet that is transmitted by a serial cable to the application on the UNIX system. The packet is then examined for contents such as extension number requested, and where the call is coming from (i.e. back from a ring no answer condition, or this is a transfer of a call directly into a voice mail box). The beauty of this is, as a software developer, and you as a purchaser, you are protected from all the intricacies and anomalies of the switch manufacturers signaling.

Although it provides a simplified buffer between the switch and the computer system, this implementation does consume extensions on the switch for the signaling.

Types of Interfaces

Interfaces describe how the computer and the telephone equipment talk to each other. As we have seen this is usually achieved by using a switch or a direct connection from the telephone company to your computer. The following reviews the most commonly found interfaces used in a computer telephony environment.

Analog

Analog interface has been already discussed at some length. The main issue to know is this type is based upon electrical impulses that are later translated into sound, which are analogous to the original speaker, or sound that was generated at the other end of the phone. This is the most basic kind of interface, and it is used for signaling between most kinds of telephone switching equipment and a computer system. The chief limitation of analog transmission is the speed of data that can be transmitted. Since all analog telephones must be able to understand how to process a limited bandwidth of sound, then there is a limited amount of data that can be pushed through that small pipe.

A related subject to analog transmission is modem theoretical upper limit of performance. The manufacturers of these devices use very advanced software algorithms to compress the data before transmission, and the sound waves being sent between them, in order to provide the current levels of data transmission. That is why digital signaling is an important resolution to this problem, since it has enough bandwidth to carry both voice and data at the same time.

Direct Inward Dialing

Direct Inward Dialing (DID), known in Canada and the USA by the trade name of Centrex, is really a description of where the telephone lines are originating and terminating, as well as being an interface. In our discussion so far we talked about a private branch exchange managing all the internal calls and outbound calls within an office. All the telephone calls come from the telephone company central office switch, and terminate on your private branch exchange (PBX), and then the calls are transferred to the extension under human or automated control.

As an alternative to purchasing and managing your own private branch exchange, you can rent from the local telephone company a block of lines directly from the central office switch which then terminate to the phones on the office desks. You can call another extension as though you had your own in-house switch, although your switching equipment could be kilometers away. You then can have all the services and support of the equipment performed by your telephone company. You may want to rent voice mail boxes from the telephone company, using a voice mail system that is usually located beside the central office switch.

If you want to have your own UNIX based system to provide voice mail, or an interactive voice response application or even a call center to provide better flexibility and customer management, then you can purchase such a system that supports direct inward dialing. Calls destined for the interactive voice response application, would travel from the central office to your UNIX computer system in your office. A specific subset of the block of telephone numbers would be assigned for that destination. While in the interactive voice response application, the customer may elect to transfer to a sales person. The call is then taken back to the central office switch (we assume here this is an unsupervised transfer), and it is told by the UNIX system to try to ring the sales person's extension back at your office. If the sales person did not answer the phone, the call is sent back to your UNIX system with a status of call failure, and the voice mail system then plays the voice mail message for the sales person, and the caller leaves a message. (Alternatively, the UNIX system could instruct to try a backup sales person).

The key to the idea above, is that your firm will then not have to manage the private branch exchange, and yet provide some more customer support that is under your direct control. The ability of the UNIX based software to interface with a central office telephone switch using direct inward dialing should have all the capability you need. Since you do not own the telephone switching equipment, you may not have access to all the features or switch programming functions you need to make your application successful. This arrangement works well in certain specific kinds of environments. The point where it makes more sense to internalize the switching functions inside your own office is to provide your firm with all the functionality you need, as well as to reduce costs.

Integrated Services Digital Network (ISDN)

Integrated Services Digital Network (ISDN) is a digital cable that has significant bandwidth. It consists of 3 channels, where one channel is used to monitor the success of the transmission of the 2 other channels. The 2 other channels are known as bearer lines or "B" channels, and they can sustain speeds of up to 64KB (Kilobits per second) each. A bit is a single binary digit that has either an on or off state. A character in a word on this page represents 8 bits put together to make a byte. Given that a page of this book is about 1900 characters of text, then a signal "B" channel could transmit 4 pages of this book in a little over 1 second. Since a top of the line telephone modem on the market today can push up to 28.8KB dependably on the cleanest of telephone lines, less than half the rate of a single "B" channel, you can see the advantage of a digital signal.

With a digital line like an ISDN line, you can mix and match what is being transmitted. Since you have 2 "B" channels, then you effectively have one channel that can be dedicated for data transmission, while the other is used for voice transmission. The physical connection between the telephone network to the UNIX system running the computer telephony application is provided by a ISDN card within the UNIX system. The application software then controls the call and data flow across the ISDN network to the service agents. This technology is a telephone company's central office service and does not require any private branch exchange equipment.

It is possible to have a distributed call center by using ISDN technology. A distributed call center is unlike a call center where there are many customer service representatives at one location. Because of ISDN, when a call arrives, the information as well as the voice is transferred to the next available representative which could be located down the block, or across town or even across the country. Even though these groups of workers are geographically apart, they can be managed from a central location and view the same centralized database of customer information. The caller will never know the difference. The immediate benefit is reduced cost of rent, since the company does not need to have a large single call center location. There is also better retention of staff, since they need not commute very far to get to work, and also a potential to reduce telephone costs. This is important to know about, since it is a medium with which other kinds of workers such as telecommuters can use successfully to help reduce organizational costs, and still provide a significant level of customer service. The installation of ISDN technology is well deployed through countries in Europe, such as Germany. Although it has not been as widely adopted in Canada and the United States, the advent of growth and interest in the Internet has spawned a renewed interest and sales of this communications technology. The result has been that the costs have come down.

➤See Case History: The "Virtual Distributed Telecommuter" - Georgia Power *on page 171* on how ISDN is used for call center applications

T1

T1 is a standard that supports up to 24 digital channels of voice or data, at a speed of 1.54 megabits per second. A single T1 can handle up to 670 calls per hour, if the average length of call is about 1.5 minutes long, with only 1 call out every 100 having the probability of receiving a busy signal. Typical applications for such large volume needs are call centers, large institutional interactive voice response systems. It is not uncommon to find a single UNIX system being used to manage up to four T1 lines for a total of 128 telephone channels at the same time. There is one case history in which the company used specialized industrialized personal computers, where up to 7 T1s have been successfully installed in the same chassis. This is where the UNIX operating system demonstrates its reliability and capability to handle many things at one time, and be able to scale to larger and larger capacities, and it is something that rival operating systems can not usually successfully manage.

The T1 digital line consists of a 2 wire pair cable that attaches to the back of your switch in your office or directly to the T1 card in the UNIX system. If the T1 cable terminates onto the T1 card in the UNIX system, then a ribbon cable connects from the T1 card to the voice processing cards. The voice processing cards are then under control of the UNIX based computer telephony software.

From a managerial point of view, the reason for using a T1 that it is usually less expensive in Canada and United States to rent from the telephone company, than renting 24 regular analog telephone lines. The cost break point depends on the tariffs in your locale. From a operational point of view you may be required to purchase a CSU/DSU device which resides on the T1 card, or it may be a stand alone device. This is a piece of testing equipment, which allows the telephone company to test the quality of the line where it terminates in your building. It also acts as a buffer between the T1 card on your UNIX system, and the telephone network, in case there is a fault in your system that may send errant signals to the telephone network.

There are even higher speed bandwidths such as T2 (6.312 megabits/second) to simultaneously support at least 96 concurrent messages, while a T3 has a capacity of 44.736 megabits/second which uses fiber optic cable and can support at least 672 concurrent messages.[12] These digital lines are used as the backbones of large telephone, and data networks.

Primary Rate Interface

Primary Rate interface is a variant of the T1 and E1, in that its signaling method is the same as the integrated digital services network (ISDN) protocol. If you were to terminate a primary rate interface line to your system, the voice processing card would have to understand the primary rate interface's integrated digital services network protocol. These are voice processing cards from various 3rd party manufacturers that are designed for this purpose, and have software device drivers for various UNIX platforms such as SCO, Sun, QNX, and others.

E1

E1 is much the same as the T1 in how it is used, but it is the European standard for achieving the same type of capabilities. Instead of 24 channels it consists of 32 channels. Hence it can support up to 895 telephone calls that last an average of 1.5 minutes in length. The functionality is very much the same as a T1, except that different countries adopt slightly different signaling handshaking functions. So it is important, when your technical person is planning to purchase an E1 card from a company such as Acculab, or Dialogic that they purchase the proper card for the destination country. This is a general rule for most voice processing cards.

Voice Processing Card Buses

A bus in this context has nothing to do with anything that has 4 wheels. A bus in the computer world refers to a specification of how data is transferred from the main memory of the computer to the peripheral equipment such as video cards, internal modems, and other devices. When the IBM PC was engineered in 1980, it was designed with an Industry Standard Architecture (ISA) bus. This specification was revolutionary in that IBM, known for keeping its technology a secret to capture its market, openly published the specification to allow any manufacturer of peripheral cards to build for the system. This is why you can purchase a myriad of peripheral devices that easily plug into the back of your Intel based system. The publication of the specification also allowed other manufacturers to produce similar computers, and a whole new industry was born. The demands for more capacity to transfer data from peripherals to main memory and back has become greater and greater over the years. As a result, newer bus technologies have evolved, such as Extended Industry Standard Interface Architecture (EISA). IBM tried to capture back some of the market with a faster bus system called Micro Channel Architecture (MCA), but it was not widely accepted. Other buses have also appeared such as Peripheral Component Interconnect (PCI), which was developed by Intel. Intel claims this bus can transfer 132MB/seccond, as compared to the 5MB/sec of the Industry Standard Architecture.[13] Many personal computers will have a mixture of buses such as Industry Standard Architecture

(ISA) and Peripheral Component Interconnect (PCI) in order to provide backward compatibility to the previously purchased hardware. In voice processing cards, the bus architectures are not limited to the data flowing between the main processor and then back to the peripherals. Special buses had to be created to handle the large data volumes that are required in voice processing. With data input in the many megabits per second, with multiple T1 cards inside the system, the personal computer bus could never move that much data and still be able to have a reasonable chance of having a clear conversation on the phone with many different simultaneous callers.

If the personal computer's bus which hosts the voice processing cards cannot physically move the data from place to place fast enough, how has this been achieved? The answer has been to host the cards in the personal computer like any other peripheral card, and also to daisy chain a ribbon cable from the top of one card to the next card. For example, a T1 line terminates into a T1 card inside the system. Its major job is to maintain the digital signaling for all 24 telephone channels and keep the data incoming, and to maintain a dialogue with the remote switch or other equipment feeding it information. A T1 card does not answer a call, it just provides the data and separates the individual channels of information. Multiple voice processing cards will reside beside the T1 card within the personal computer. A ribbon cable is chained card to card from the T1 card to the voice processing card. In such a configuration, and depending upon the voice processing card manufacturer, there could be as many as 1 to 3 cards needed so the calls can be answered. There could be more voice processing cards if there are other interfaces or special resources, such as fax, text-to-speech, or speech recognition. Open a cabinet of a personal computer so configured, makes for a busy looking system crammed with cards. Specialized manufacturers have emerged to meet this market, where they supply systems with 15 to 20 slots in the back of a single "industrialized" personal computer. As components and more computing cycles can be embedded on the cards using more powerful digital signal processors (DSP), the more voice processing power is at hand. As a result fewer number of cards are required to be populated within the chassis of the computer.

The buses that have evolved in the computer telephony marketplace are PEB, MVIP, and SCSA. Dialogic Corporation, which is the largest manufacturer of voice processing cards in the market, introduced the Pulse Code Modulation Expansion Bus (PEB). Not to be too technical, pulse code modulation refers to the ability to convert analog signals (such as speech) into a digital stream of data. The expansion bus determines how the data is being transmitted along the ribbon cable. Dialogic proposed this as an architecture for its own products. Although they have a dominant portion of the market, the market wanted an open architecture. Dialogic did finally publish the PEB specification to allow other competitors and manufacturers to design cards to work with this bus. Other manufacturers saw this as an opportunity to provide an alternative bus for voice processing, and another specification called Multi-Vendor Integrated Protocol (MVIP) . The technology was developed by Mitel, a switch manufacturer, and the standard was proposed by Natural Microsystems, one of Dialogic's competitors. It has been accepted widely by other manufacturers, and the marketplace. This leaves the business person with a choice, but one can not usually mix and match the card technology within the same system, unless the card manufacturer supports both bus structures. These bus structures are limited to what can be achieved in a single personal computer.

In order to alleviate this, Dialogic proposed an open standard called Signal Computing System Architecture (SCSA). This was announced in the Spring of 1993, with wide industry support both from hardware and software vendors. It provides for a new way of building very large and multiple systems to be integrated into one virtual system. It allows other manufacturers of cards to build cards which will work with other card manufacturers such as Dialogic, Natural Microsystems, and others. The Signal Computing System Architecture is as significant to the computer telephony industry as was the publication of the Industry Standard Architecture in 1981 by IBM. This provides a software and hardware model that is open and standardized.[14]

By the way, there is one manufacturer, Amtelco, which produces voice processing cards which have both MVIP, and SCSA connectors. This allows previous cards that were MVIP based to now work with the newer SCSA based cards. As a business person, (who I hope I have not bored with all this technical background), you need to know about this so you can make a decision on your investment in the equipment you are planning to use. You can select products that are standardized on the bus structures, and be aware the decision will effect your ability to grow and build upon your foundation of technology for the future.

Database

A keystone of many computer telephony applications is the corporate database, since that is where all the customer information resides. Remember, computer telephony helps the customer. In this section, we will do an overview of database technology. This review is not meant to be a detailed text book review, and there are many good books on the subject. The purpose of this is to give a sense of how the data can be organized and how the information is retrieved. The database is the central tool to understand for most computer telephony applications.

Database in the Corporation

What is a definition of a database? A database is an organization of information in a structured way, so that the information becomes a central common resource for the organization. This also implies that many people can simultaneously access the data. Users obtain different views of the database information, which would be near impossible to perceive in a manual paper system. There are different kinds of databases that have been developed, which are relational, hierarchical, and network. The hierarchical database structures its data in an organizational tree like top-down structure. The network database organizes data in a circular fashion, where individual records with common associated properties are organized into cells, and each record and cell has the potential to reference each other. These kinds of databases, especially the network database can retrieve data in a very fast manner. They both have a drawback. Once they have been designed, it is difficult to change how they organize their data or be able to add new record structures to them.

Conversely, relational databases can be changed quite easily by the database administrator, where they can add new tables and generate new screens and reports. There was a small trade off in the early days of relational databases. Although at first they traded speed for flexibility, they are the predominate database systems used in business today. Relational database technology has surpassed the other database types in usage. Relational databases typically used for the UNIX marketplace are Informix, Oracle, Sybase, and Progress. These database products are sophisticated in that they can have networked computers located in different parts of the world, and be able to have users create queries against the database tables to produce reports.

What was it like before databases? Let me use a university example to guide you. The university's business office would usually have student contact and billing information. The university registrar office would normally deal with the student contact and details on the courses taken, and degree program being earned. The Dean's Office needs to maintain current professor contact, remuneration, tenure status and class schedules for each department, and within each department, schedules for each professor.

Before database technology really took hold in the late 1970's and early 1980's, programmers used programming languages such as COBOL to build specific applications with their own set of data. They would build a Dean's Office application, and then they would build the Registration Office application, and finally a Business Office application. For example the student contact information is shared by at least 2 departments, and possibly a third. Do you see other things in common between the departments?

The data for each of these applications was usually stored in an isolated manner. Yes, there was some sharing of data between applications, but the data concurrency became very unmanageable at times - especially for large organizations. If it was at all possible to share data stored in different files between applications, then the concurrency of the data had to be synchronized. In other words, certain programs had to be built and run at specific schedules so that data in one file could be used to update other files, to allow other programs to work. This forced programmers to become more concerned in the management of the data than trying to solve the business problem at hand, such as producing course schedules, or reporting students who are late in paying their tuition.

What resulted was a lot of programmers and system administrators writing a lot of programs that had to be changed and maintained again and again. Even worse, the programs were organized and written in such a way that they were very closely organized logically, and structured as to how the data was organized in the files. This meant there was no data independence between the programs and the actual data stored in the computer files. If a new business requirement or registration requirement was needed to be added, changed or deleted in the system, then this sometimes created an expensive domino effect. A change to just one program or data file may require many changes to the many different programs and data files to accommodate the new request. All these changes created inevitable human error in program logic, which meant that all facets of the system had to be tested. All manner of possible combinations of bugs in the logic had to be eliminated in order to make things work. In other words, the programming department could not keep up with the demands of the organizations needs

I fondly remember these days in the early 1980s, when the expense of the database technology was much higher than it is today. At that time in my career, I was a entry level programmer, and I would witness systems analysts who would purposefully manipulate or withhold information given to their end users (read customers). They did this to minimize the changes they had to do, in order to lower their MIS requirements. Occasionally, this became a political tussle between the managers who ran the business and the Management Information Services organization trying to keep a lid on the demands that were being made of them. At times, this was not a good compromise.

Then the personal computer revolution created a power shift. It was 1981, and I distinctly remember watching a summer student on staff unpacking the first IBM PCs running DOS 1.0, and asking myself, what is this wonderful device? As this technology has grown and developed, it has allowed corporate workers to manage their own data, and their own queries against the corporate database. This is why the personal computer revolution has turned the tables on the system analysts and MIS managers of yore. Some would argue that the personal computer revolution has allowed too much freedom and capability to those who can guillotine themselves most easily and efficiently. Users have a lot of power. It has to be developed, nurtured and appropriately used to make things work out not only for the individual, but for the company as a whole.

Database of Databases - Data Warehousing

Essentially a database is an electronic representation of things in the real world, and how they are related to each other. Database technology provides the ability to remove redundant data, and as well, allow management of data in such a way that different users can have simultaneous views of the same data source. From the MIS department point of view, database technology has evolved to eliminate or reduce the work demands, since users have the tools at their own desks to manipulate data. It provides data independence so that they can eliminate the need for large changes in their programs. Recently, database technology has gone beyond the day to day mundane activities of just accepting data and printing or displaying information in an organized or ad hoc manner, to manage daily business functions. Because of the tremendous reduction in the cost of hard drive storage and low cost computer resources, companies with vast amounts of customer data can now do very accurate statistical trend analysis that were impossible in the past. This trend analysis is the basis to help predict what products or services will be most likely be sold in the future, or what areas of niche business areas not known to management beforehand now should be exploited to increase profits.

The term for this kind of database and analysis work is called data warehousing. Data warehousing can be found in call center applications. If there is a new trend in the data that can be exploited in a sales and marketing campaign, then a predictive dialer may be used in conjunction with the subset of the customer base, to gain new market share or exploit a previously undiscovered facet in the market. A data warehouse is a database of the databases in the corporation. In large organizations that have customized applications developed by different departments over time, the result may be that the call center agent may have a screen full of applications to manage. Each application is working with its own set of databases, and could also be residing on different computers. This is a hindrance to the call center agents, since they need to enter the customer's information over and over again within each application. This takes up valuable customer time. To resolve this predicament there are 2 available options. The first is to look at purchasing an off the shelf product that is already designed for call center applications that use an open database system. This would be able to talk to the legacy databases, yet provide a common new and multiplatform interface to gain a single point of access to the customer information. Such a company is Scopus, which is reviewed later under the chapter on call centers. The other solution is to use a data warehouse which will allow for the centralization of the customer data, and a custom application that unifies the inquiry process.

Introduction to SQL

Let us review a hypothetical database for the university we have been talking about. This kind of database will most likely be much more complex than what we will describe here.

Databases work so that the programs and most importantly programmers and end users do not need to know how to actually do the detailed search for the data. This is achieved by saying to a database program 'For student Joanna Dobson list the tuition owed." This type of request is expressed in a database language, of which the most popular is structured query language (SQL). The language is syntax that is executed within a program, which then retrieves the information to be listed on a screen or report. By having a flexible database query language, this provides data independence. Before languages such as SQL were developed, programmers had to reserve program memory within there programs to match the structure of how the data was stored in the data files. This is very labor intensive. If one field changed, then possibly many programs would have to be re-tested.

The request above is not expressed in a SQL statement. We will show you how the data is stored, and what the request is to do that, using a SQL statement. SQL statements are used in a relational database. Let's be a database administrator for our hypothetical university. You have to organize the data for the request above. You may think that a database administrator would be most concerned about how to store the data into the database. Quite the contrary. A database administrator designs his database by how it is going to be outputted to screens or reports.

➤If you hunger for more complex SQL examples, please see Appendix C on page 338.

The request being asked is:

"Give me the tuition owed for a student called Joanna Dobson".

The report being generated must show her contact information, the course names, the professors instructing the class with date and times, and location of the courses.

In a relational database information is stored in tables. Each row in the table represents one entity, while the columns describe attributes of that entity. See Figure 1-5 Database: University Student Table on page 53.

Student ID	Last Name	First Name	Street	City	Prov.	Nation	Tuition owed	Phone Number
78911373	Kincaide	John	149 Hazel Street	Waterloo	Ontario	Canada	20,000	519-555-8753
20063731	Dobson	Joanna	123 Louise St.	New Lowell	Ontario	Canada	0	705-555-2112

Figure 1-5 Database: University Student Table

To list the Student First Name and Last Name for a particular Student ID, the SQL statement would be:

SELECT Last_Name, First_Name, Tuition_owed from Student
WHERE Student_ID="2006373"

Output would be:

Last_Name	First_Name	Tuition_owed
Dobson	Joanna	$0

Figure 1-6 Database: University: SQL statement and result for
searching for a particular student

Database Use in Computer Telephony Applications

Database as previously mentioned is the cornerstone of the
computer telephony environment. As well, the UNIX platform has
been seen as a high volume and reliable platform to host this kind
of application. We will examine the typical usage of databases in
different kinds of computer telephony applications.

INTERACTIVE VOICE RESPONSE

In interactive voice response applications, such as banking by
phone or student registration systems, the telephone becomes the
data entry device, instead of a personal computer or simple
terminal. Instead of forms to fill out on a screen, the caller is asked
questions to answer using a touch tone phone. In order for a
student to know the outstanding balance of fees owed in our
example above, the student would call and be prompted for their
student number. In this case the student would key in "20063731"
and the interactive voice response software would insert those
numbers into the SQL statement above, and execute it. The
database would return the value of FEES OWED as zero, and
then say to the student that they do not owe the university any
money. The connection between the telephone and the database is
performed by a voice processing card in concert with the
interactive voice response software. Interactive voice response is
usually the simplest method of providing customer access to their
information from an installation point of view. It directly connects
the caller to the database.

CALL CENTERS & PREDICTIVE DIALERS

Call Centers use the database in a more expansive way. The call itself, using caller-id or prompts from an interactive voice response system, spawns a database transaction that is dynamically displayed on a screen full of information for the agent. This means the agent does not have to key in customer information, and speeds up the whole process.

Predictive dialers are used for outbound call handling. A predictive dialer eliminates the manual dialing of numbers. Once a person answers the phone, (i.e. not an answering machine, or no answer, or busy), then the called party is immediately connected to a customer service agent, and the data on that person is presented on the agent's screen. A database of phone numbers and customers is loaded into the system. Call centers sometimes "blend" both inbound calls with a predictive dialer, to maximize the quality time the agents spend talking to prospects or customers, while minimizing customer wait time.

Background on Speaker Recognition

Speaker independent recognition is the technology that allows a speaker of a given language, such as English, or French, to command an application with verbal instructions. The use of this technology has applications within many computing applications, of which interactive voice response is one. We need to quickly review what speaker recognition is, and what are the differences between speaker independent recognition and Speaker Dependent Recognition. Then we will discuss discrete, and large vocabulary speaker independent recognition.

Speaker Dependent Recognition

Speaker dependent recognition requires the person using the software to 'train' the software to learn the unique speaking traits of the speaker. Depending on the implementation of speaker dependent recognition software, the user will read aloud a couple of paragraphs of a nonsense story. This is spoken aloud two or three times to the computer. Language and speech are comprised of different building blocks. The principle parts of speech that effect speaker recognition are called phonemes and consonants. The nonsense story that the speaker reads to the computer is comprised of as many combinations of such parts of speech as needed for the software's 'recognizer'. Each time the speaker reads the story, the software 'learns' more on how the user speaks the parts of speech, and is able to recognize the speaker. This is the type of speaker recognition being sold to provide basic dictation systems on high end personal computers. By no means, is this error free, nor can you speak to the computer at a high rate of speed, and not in a office which has a lot of background noise. You have to speak the words slowly enough for the syllables to be distinguishable. For example, the word 'automatically' would have to be pronounced 'auto-mat-ic-ally', with slow cadence between the syllables.

Speaker Independent Recognition

Speaker independent recognition does not require the software to learn the uniqueness of the person's speech patterns. This would be impossible in computer telephony applications. It is not practical to have millions upon millions of unique speech patterns trained, stored and accessible anywhere in the world over the public telephone network. In order to solve this vexing problem, research into speaker independent recognition has taken two avenues of approach, and has generally provided two types of products. The two types are, discrete and large vocabulary speaker independent recognition.

DISCRETE SPEAKER INDEPENDENT RECOGNITION

The first kind is called *discrete* speaker independent recognition. This kind of speech recognition is usually restricted to a series of recognizable zero through to nine, and some key words, such as 'operator' or 'help' or 'exit'. The ability of the software to recognize this limited list is usually very high. This kind of speaker independent recognition uses a statistical model of speech for a specific population of speakers. English as a language has many speakers who have completely different types of accents. There are noticeable differences between speakers from United Kingdom, Canada, Australia and the United States of America. Within the

United States there are marked regional accents. A person from New England will sound different versus a person from the southern USA. (While I was traveling in the southern USA last year, a hotel front desk clerk with a very heavy southern USA accent, mistakenly thought I was from the United Kingdom, because of my central Canadian accent. I do not have an UK accent at all.) Even within New England, a person will have different accents depending whether you are from Maine, New York, or Boston.

How is a poor computer software program going to cope with all this variance within a single language? The statistical model analyzes a distribution of speech characteristics for a small set of words within a specific regional voice sample. What researchers and software developers of this kind of speaker independent recognition do, is to sample as many as a 1000 or more people saying the same words like 'zero', 'one', 'two' etc. This builds up a distribution or representative sample of how a speaker sounds from that region of the country. The parts of speech are analyzed for this population. This is one of many reasons why you will see vendors of speaker independent recognition advertising products for United States English and United Kingdom English. This distribution is then statistically modeled, so when a caller is asked to say "eight" in response to an interactive voice response prompt, it is able to compare that caller's word to the representative samples it has on file. A language is not a static entity. It lives and grows every year, and slight changes in pronunciation of words occur over time. Like other statistical modeling test methods, the sampling process and recalculation of the distribution model has to be updated on a regular schedule.

interactive voice response System: "Please speak the 4 digits of your customer account number individually after each tone - BEEP".

You reply, "Nine".

The system accepts that input, and then prompts for the next digit -"BEEP".

You reply "Five".

The system does another "BEEP".

You reply "Three".

The system prompts again for the final digit "BEEP".

And then you say "One".

Figure 1-7 Example Dialogue of Discrete Speaker Independent Recognition

As well, the word 'eight' has to be captured in such a way as to reduce any other influences, so the match can be made between it and the samples on file. So this is where the adjective <u>discrete</u> applies. When a caller is prompted, they may hear the following while being asked to enter their 4 digit customer number in response to an interactive voice response prompt.

It works, but imagine if you had to speak a 16 digit credit card number! Those beeps would drive you crazy! If the software did not quite recognize one of the digits, and could not verify the credit card number, the software would ask you to do it all over again. For some people, this is not a very pleasurable experience. The 'BEEP' prompt by the software allows for distinctive break up of the individual string of numbers being announced to the software. This break up of the numeric string gives the software program a better chance of understanding the meaning of the individual word.

What are the technological and commercial reasons for using this type of speaker recognition? Technically, not all of the United States and Canada have touch-tone service on all telephone company central office public branch exchanges where the telephone interfaces. This tends to be the case in established older rural areas of these countries. This means that callers are calling on a rotary phone that uses 'pulse' dialing to determine the digits. In some other countries in the world where there is very little in the way of touch-tone service, the only way to communicate is by using the older pulse dialing. There are only two choices to get around this problem. One methodology costs more to implement than does the other.

As the owner of the interactive voice response system you could purchase a pulse-to-tone converter. This highly technical hardware listens for the 'click' pulses on the line, and then converts them to equivalent touch-tone tones. These devices tend to be expensive and need a skilled, expensive technician to tune the device to local telephone environmental conditions. They are also susceptible to background or involuntary noise. Events such as crackling static on the telephone line, or even if the caller had a bad hacking smokers cough, the pulse-to-tone converter may interpret these sounds into unintentional touch-tone digits. This would send the caller into a part of the interactive voice response application that they or you did not want them to access.

The other solution is to use discrete speaker independent recognition to announce the digits, and potentially provide a higher level of accuracy and control in such situations. This kind of speaker independent recognition is used as an adjunct methodology to an existing interactive voice response application that is expecting to receive numeric strings of touch-tone digits. As you will see from the imaginary doctor calling scenario in the case history The "Natural Advantage" - Parlance Corporation's Name Connector on page 89 in the chapter on interactive voice response, this still forces the caller to interact with the technology, and follow the demands of the limitations of that technology, rather than having the technology fit the needs of the caller.

LARGE VOCABULARY CONTINUOUS SPEAKER INDEPENDENT RECOGNITION

Large vocabulary continuous speaker independent recognition does not use a statistical model to represent a portion of speech, but rather is based upon the constructs of speech itself.

Primarily, this facet of the technology has focused upon the phonemes of speech. It has been found that phonemes are the sounds that are the basic building blocks of a language. Each language has a well defined set of phonemes and other constructs, which are independent of the idiosyncrasies of most speakers within that language. Speaker independent software of this type must analyze the sound wave patterns within the speech, so as to recognize the phoneme constructs within the sound wave pattern of the speaker. As a result, this type of speaker recognition does not require the "BEEPS" to aid in the recognition of the words, as does discrete speaker independent recognition. As well, another key factor is that the speaker independent recognition has a large vocabulary of words. Each word in the vocabulary is analyzed and broken down by its phonetic components into a phonetic database. When a speaker talks to the system through the telephone, this person's phonetics and how they are arranged within the words he or she is speaking, is compared to the phonetic database stored on the system. At the time of this writing, continuous large vocabulary speaker independent recognition can usually handle immediate demands for up to 2000 words at any one time, with a potential vocabulary database that can hold 40,000 words. There are technical reasons why 5% of the database can be managed at any one given moment.

BBN's Major Innovations

- Context-dependent phonetic HMMs (Hidden Markov Models) (1984)
- Forward-Backward search (1986)
- Rapid speaker adaptation (1987)
- HMMs for word spotting (1987)
- Detection of new words in running speech (1989)
- N-Best for integration with natural language (1989)
- Real-time spoken language on workstation (1990)
- software only on SGI (1992)
- real-time, continuous speech recognition, 20K words (1993)
- real-time, continuous speech recognition, 40K works (1994)

Figure 1-8 BBN: Major Innovations[15]

One may think that this is more than enough words to develop complex dialogues with the computer system. You may start to think that you could dictate your own letters, and also request and perform a large number of tasks through this kind of automation. Well, don't plan to reduce jobs at your company just yet. Although this is the most promising form of speaker independent recognition, it still has limitations of understanding. The limitations occur in the ability to handle many different variations of requests. If the requests are simple and direct, then the software can handle the situation relatively easily. Where it falls down is when the caller wants to make a lot of changes, or needs new and varied preferences to his or her requests. The software requires a finite set of words, and a finite set of natural language understanding to be able to cope with verbal requests.

| BBN's Key Technology Milestones | |
20 Years of Experience	
1976	HWIM, one of the earliest continuous speech understanding systems.
1983	Real-time continuous speech recognition system developed for a US Navy TRIO F14 Radar Intercept Officer training application. The 100-word vocabulary system ran on a VAX 11/780 with an FPS array processor.
1987	Near real-time demonstration of the BBN Byblos continuous speech recognition system. The 1,000-word vocabulary system operated on a 128-processor Butterfly.
1990	Real-time operation of the Byblos system on a Sun 4/330 workstation using an additional DSP board.
1992	BBN advanced continuous speech recognition product. Operates in real-time on low-cost UNIX workstations without additional accelerator hardware. Integrated speech recognition and natural language understanding on UNIX workstations.
1993	Demonstration of 20,000 active word, speaker-independent, continuous speech recognition in real-time.
1994	Further algorithm improvements increase continuous dictation vocabulary capability to 40,000 words
Figure 1-9 BBN: Key Technology Milestones[16]	

The BBN's major innovations over the years are recounted below. As you can see the innovations are based upon new software algorithms to manage and recognize the speech. These innovations have allowed for better recognition of the words being spoken, and are able to search for and understand the words in the context of what has been spoken.

The research into this area by BBN goes back many years, and as shown in "Figure 1-9 BBN: Key Technology Milestones" on page 61 it has traversed many computer systems in its evolution. As you can see from a hardware perspective, the more computing power available the more speech recognition capability. As well, you will notice in their implementation that they were able to use Digital Signal Processor (DSP) technology in 1990, as an adjunct specialized source of computing power to work the software algorithms. Then they were able to use Silicon Graphics Inc. UNIX workstations using the very powerful MIPS R3000 and later R4000, as well as Intel Pentium central processors, to provide the necessary computation resources for the software. In doing so they have been able to increase the capability of the software each time.

Most personal computers being used for speech recognition for computer telephony applications, will have a voice processing add-on board that uses digital signal processor technology, to provide the horsepower to run speech recognition software. Usually, the voice cards with the digital signal processors that are used for speech recognition are a shared resource for all the potential telephone lines, coming into the computer. For example, a computer may have 8 or 12 telephone lines with an extra voice card with the speech recognition resource, that will support 4 or 8 simultaneous requests for speech recognition. This works well, until there are more simultaneous demands for the speech recognition than the resource can provide. Such configurations meet the needs of most applications, but the goal to have all telephone channels having the speaker independent recognition available on all ports is obviously much better. Moreover, the shared resource concept also makes demands on the hardware that must be put inside the chassis of the computer.

The special speech recognition board occupies a slot in the back of the computer. If more shared resources are needed, then more boards have to be placed in the computer system. This is great for the voice board manufacturer's profitability. For the owner of the system, it can become an expensive headache to purchase and manage this resource. The irony of this situation is that if the cost could be lowered and the functionality of the product could be increased in a software only solution, then the overall acceptance of the technology would be greater. This would mean that the income from the sales of such software would then fuel more research and development, to provide more functional product for general use in our daily lives. It is almost a vicious cycle. It seems that with the right hardware and operating system, with the best speaker independent recognition software, the technology would be more useful.

It will come to pass, because the compute resources and the operating systems such as UNIX, and Windows NT on Intel Pentium and Pentium Pro central processing units and other such powerful chip sets, will finally allow this technology to really blossom in the next couple of years. With the development of the 64 bit (very, very large computational resources) chip sets and most importantly a 64 bit UNIX operating system code name "Gemini" destined for shipment in late 1997 or early 1998, as a joint venture of the SCO, Hewlett Packard, and Novell, the issues surrounding computing resources should be very much minimized. Now the product of all these years of labour will be available to a mass market, which will lower the cost of the systems over time. Most interactive voice response systems primarily use voice and fax functionality alone or in combination with each other. The significant added value that large vocabulary speaker independent recognition will be realized.

Generally, there are two key elements that allow continuous speaker independent recognition technology to manage hardware compute resources more efficiently and at less cost. First of all, since the system is using powerful central processors, that are now relatively inexpensive, there is no need for additional digital signal processor hardware. This allows this technology to be a regular software only solution. The second element is that they are using the UNIX based operating system which lends itself to process more than one thing at a time, since it is multitasking, to help analyze the speech process.

Speech Reproduction in Computer Telephony

Speech reproduction is achieved using two methods. One method is speech concatenation, while another is text-to-speech. We will review the reasons why and when the two methodologies are used. The first uses recorded speech, while the latter is the opposite to speech recognition technology. Text-to-speech technology takes strings of text and numbers stored within the computer, and converts it to speech.

Concatenated Speech

Concatenated speech uses professionally recorded phrases. These phrases would include such things as frequently used words, and numbers. For example, an interactive voice response system may tell a caller how much money they have in their chequing account. Let us say the account has $536. When the customer calls into the system, it will report "<The balance of your chequing account is..">, <very short pause> "<Five hundred>", <very short pause> "<Thirty-six>" <very short pause> "<dollars>", <very short pause> "<zero cents>". When the bank-by-phone system accessed the bank's database which returned the value of $536 from the chequing account, the program then strung the pre-recorded phrases together to generate the sentence. Although this provides the best quality sound as compared with text-to-speech, concatenated speech has practical limitations.

◄ Please see *Notation used in this book* on *roman numeral page xv* for an explanation of the use of "<>" and "{}" symbols within the text

Concatenated speech is expensive to have a professional voice recorded. To do a large application with hundreds of phrases takes time (possibly a couple of days), and as well, may be charged as much as 25 cents per word to record. (Some companies hire out professional voices by the hour or by the job.) Given that a system may have thousands of words, the professional voice can make a fair bit of *your* money. After the recordings have been completed, then a special editing process must take place. The individual words have 'dead air' preceding, and trailing the recorded phrase. These have to carefully edited out, otherwise the "<very short pause>" becomes a staccato of "<pregnant pauses>". This becomes tiresome for the callers on the application. If your callers are calling on a toll-free line for which you pay a flat fee per minute, then these "<pregnant pauses>" will cost you money.

Overall tone of voice and quality of the speaker makes a big impact on corporate image. You may think this is trivial, but in reality it can have quite a dramatic psychological effect on the callers' experience in using the software. For example, a bank I use has a bank-by-phone system. The voice recorded in that application sounds *too* business like. It gives you the impression that the person who recorded the application phrases was an old fashioned, hard nosed, rotund elder silver hair banker. You know the stereotype. The one who wears three piece pinstripe suits, has a gold pocket watch and chain, walks with a silver handled cane, and peers nastily through a monocle, while reviewing your business plan with an ever present scowl. When you press "*" to leave the bank-by-phone system, it says with an impatient, rough regimental sergeant-major ordering voice "Thank-you Good-bye!". The funny thing is, because this system is so useful and convenient, I happily pay monthly fees for this abuse! (No, I'm not a masochist!) Please note, the bank is saving a bundle because they do not require a human to answer these questions! By the by, I do not hold any grudge against banks and bankers. Although a human professional speaker is better than text-to-speech technology in sound quality, the emotional presentation by the human speaker can reduce the overall quality of a concatenated speech based interactive voice application.

Text-to-Speech Technology

The most common reasons for using text-to-speech are:

- The information to be reported over the phone changes frequently. For example, this means it would be impossible, even for an army of professional speakers, to record all the phrases in an interactive voice response system, that played back a news wire service stories that changed every 5 to 10 minutes.

- There is a great amount of unique information. Another example of text-to-speech use is for an order entry interactive voice response application for a distributor, or manufacturing company. There could be tens of thousands of parts (i.e. parts for an airplane or automobile). Without text-to-speech, each part would need to have the description of the part professionally recorded. It is possible to record the price, yet the outcome would sound odd when the description is text-to-speech generated, and the price is professionally recorded and played back. If a customer was asked to enter a part number on the keypad of the telephone, then the interactive voice response system would find the part in the database, convert the text description into speech, "{left aileron for Cessna 172}", and then the price would be in a professional voice. "<two thousand, nine hundred and fifty seven dollars>". The description of the airplane part would be in synthetic voice, while the price would be in professional voice. This is very jarring to the listener. In order to make the sound of the application consistent, it would be better that the part description and the price be all placed in a standard set sentence. "{Part number 82796 consisting of a left aileron for Cessna 172 for quantity one, the price is two thousand, nine hundred and fifty seven dollars}". Moreover, recording the prices is not practical because they would inevitably change over time, and would require frequent recording sessions.

Text-to-speech software analyses the individual words within the textual string of characters, breaking them down into the parts of speech. Then the whole sentence is analyzed for context, and inflection. For example if an "!" or "?" ends the sentence then the sentence would be spoken with the tone of voice that respectively commands attention or asks a question. As well, the software has to be able to handle special contractions such as "Dr. Smithwicks" to say "{Doctor Smithwicks}", and also special strings of numbers or other conditions. For example, the rules that the text-to-speech uses may see "123" as "{one-hundred and twenty-three}" or in a different context, it may be required to count the string of numbers individually such as "{one, two, three}". As well, if the sentence structure is long and complex, then the text-to-speech software has to analyze the whole sentence in one go, before starting to speak. This sometimes creates small pauses on the phone before the next sentence is spoken, because it takes time to process longer strings of words. As you can see this software is indeed complex. This is truly synthetic speech - and most of the time it sounds synthetic. Depending of the software company that developed the text-to-speech algorithm software, there is a variance in the quality of the generated speech. Generally, the unwanted characteristics are the slurring of words, or improper inflection or monotone sounds of the voice. Current generations of text-to-speech software can control the sound and minimize some of the artificial sound of the voice by changing the cadence (speed and pacing between words), and the pitch of the voice. Recent advances in this technology are growing as fast as the development of speaker recognition. Recent demonstrations by vendors like BestSpeech by Berkeley Speech Technologies in the USA of their female version of their English text-to-speech products, produces close to human sounding speech.

TEXT-TO-SPEECH UNLOCKS THE SECRETS OF THE UNIVERSE

Text-to-speech technology has many applications outside the field of computer telephony. If you have seen any science television programs on astrophysics featuring Dr. Stephen Hawking of Cambridge, England, then you will know this technology has greatly increased the capability for the physically challenged. Dr. Hawking suffers from a degenerative neuromuscular disorder called ALS. When Dr. Hawking contracted pneumonia in 1985, he had a tracheotomy operation in order to help him breath, but he lost his ability to speak. Before the tracheotomy the ALS had reduced Stephen's voice to just a whisper. Reputed as being the successor and leading thinker to Dr. Albert Einstein in the study of how the universe works, he has been able to enlighten the world of science by manipulating a small computer attached to his wheelchair to select words from lists to construct sentences of text. He has very little hand movement, but he is able to manipulate a device that acts like computer cursor keys. These sentences are then converted to speech by a text-to-speech conversion. In this way he is able to provide lectures, write books, and work with others like he had never been able for many years. Walt Woltosz, of Words Plus Inc, in conjunction with a speech synthesizer from Speech Plus both of Sunnyvale, California, learned about Dr. Hawking's troubles and they were able to provide him the equipment and training he needed. The world knows more about the universe and how it behaves, because of text-to-speech technology.[17]

Interactive Voice Response

The purpose of this section is to tell you what this is, and a little on how it works. You should review this section and the previous section on database in order to get a real handle on this portion of the technology. We will not belabor the issues here because the chapter on interactive voice response starting on page 75 providing you with detailed implementations of the technology. Interactive voice response has already been discussed in the sections on database, and speech recognition.

Interactive voice response enables a telephone and/or a fax machine with a telephone, to become a computer device connected into the computer system. The call is answered by the interactive voice response software which controls a voice processing card. It then prompts the caller for numeric and even alphanumeric information to be input using the telephone keypad. The interactive voice response software then can retrieve, change or delete information from a database or other data source. It can speak back the information using concatenated speech, or text-to-speech technology. It also can send information via fax. Delivery can be done by two methods. The two-call method asks the caller to input the destination fax machine phone number and then hangs up. The interactive voice response software then calls upon the fax server software to send the fax on a second call. The other method is called one-call fax delivery. If the caller is calling from a fax machine that has a telephone handset, the interactive voice response software answers the call in the usual manner. Instead of entering a fax number, the caller is asked to press the "start key" on the fax machine. This causes the fax machine to emit its fax tone pulses. This tells the interactive voice response software's fax module to start transmitting the fax. The benefit to the owner using one-call fax delivery is, they do not incur any long distance charges. The voice processing card hardware must have a fax hardware resource and software.

More sophisticated technologies such as text-to-speech, and speaker recognition enable interactive voice response software to provide more variable information, and accept full spoken words as input. Some applications have information that is constantly changing in the database, and text-to-speech is the only way to deliver the information. Speaker recognition software allows callers to speak words, or key phrases instead of trying to use the keypad on the phone to spell alphabetic characters. There are limitations to speaker recognition, which have been already covered.

Interactive voice response is most likely the easiest kind of computer telephony technology to deploy, and also has the ability to field from 15% to 95% of the calls that customers ask a person in the business. The usual number reported is around 15% to 30%. Can you imagine what you could do with an availability of 15% more labor in your customer service department? You could call more customers to find out what they desire to hone your products and marketing programs.

Call Center Applications - "Screen Pops" and Predictive Dialers

Call center applications provide the ability to maximize the time spent by customer service agents dealing with customer's needs or problems, rather than being tied up in administrative information collection. These types of applications demonstrate the value of computer telephony integration.

A typical call center would have these basic elements in the call flow. The caller contacts the company, and if available, caller-id is captured on the line by an interactive voice response system, or other software that is able to trap this information. If caller-id is not available, then the caller is prompted to enter a number that identifies themselves, such as credit card number, or telephone number. This information is used to find the customer's record in a database. If the customer record is found, the call is transferred to a private branch exchange with automatic call distribution software, or a dedicated automatic call distribution (ACD) system. The customer's key data is held in association with the call in a database table. Once a free agent with the appropriate skill sets (such as being Spanish speaking) becomes available, the call is transferred to that agent, and a database application is also executed at the same time. The database application takes the customer's key information, builds a screen of information, and then cross tabulating the agent's terminal location in the database, refreshes the screen of the agent with the customer's information. The agent then has all the information needed. They do not need to ask the caller for further identification, unless it is for security reasons. This is called a "screen pop", because the screen initially refreshes or "pops" without an agent's interaction. If the caller has a problem that a supervisor with more authority needs to review, the data and information about the customer is not lost when the call is transferred to the supervisor. The caller does not have to repeat his account information, and possibly the issues that need the extra authority. This not only saves time, but also the customer's need to explain the same things, over and over, to different people within the organization.

Some of the case histories indicate that 20 or more seconds are saved for each call an agent must process. In environments such as large call centers for credit card companies, which have potentially hundreds of agents, and processing millions of calls per month, this can reduce the number of agents required to look after customers. Every second shaved translates to much lower costs in people, equipment and supporting infrastructure. Call centers experience many peaks and valleys of influx of calls every day. This forces managers of call centers to hire part-time workers for different peaks, and manage the different number of regular workers for the other times. Some companies have found that by using screen pop technology they do not need to hire as many part time workers to cover the peaks, and are more accurately able to staff the regular workers. Even if you had a smaller call center of about 20 people, this time saving would still have an impact on the bottom line.

During the slack inbound call volumes, a call center manager may assign some of their staff to do outbound dialing tasks to follow-up on good customers, and if necessary call those who have not been timely with their payments. They have a list of clients on their screen. They pick up the phone. They call the prospect or customer and get a busy signal, or an answering machine, or there is no answer. They finally get a customer, ask them to enter the customer's identification, and get a screen full of information, and can talk to the customer about what they are offering or when they are paying. This can be a labor intensive task. It has been shown that in one hour, about 20 minutes will be spent talking to a customer, or prospect, and the other 40 minutes will be spent manually dialing.

Would you like to increase their "talk time" by almost 300%, to about 55 minutes, of your customer service department staff? Another technology that is related to screen pop is predictive dialing. Instead of dialing manually, a predictive dialer is software that will dial the numbers for your staff, and once it detects a human voice that is real, (as opposed to an answering machine), a screen pop occurs with the customer or prospects information. The predictive dialer is loaded with customer numbers from the corporate database. Each list of phone numbers would be associated with a specific task. Each task may require a special set of skills. These skills are identified, and the customer service agent's profile may have these skills entered in the database. When the call volume drops to a predetermined level, the predictive dialer will start dialing numbers, and then present the agents with outbound calls instead of inbound calls. The outbound calls are identified by the nature or identification on the screen. When a predictive dialer is used in this outbound and inbound call environment, it is called "call blending". Predictive dialers are usually used in telemarketing organizations to do only outbound calling.

You can see how an interactive voice response system, inbound call screen pops, and predictive dialing screen pops provide the degree of automation to increase sales, and lower costs, and provide greater customer service. One of the challenges in the market has been providing seamless as possible integration between these different kinds of products, and be able to handle the interfaces to the proprietary nature of the private branch exchange systems. This is a need which in the past 2 or 3 years has been addressed by companies such as Dialogic, with their CT-Connect, Aurora Systems, and Genesys Labs with their T-Server software. These products are called "middleware" and they provide the interface and glue between the applications such as interactive voice response and database applications that use screen pop technology for both inbound and outbound calling. Some manufacturers of private branch exchanges provide all this in a "soup to nuts" approach. They will sell you the switch, and then provide a proprietary interface from which you can program your database applications to do the screen pops. This is the most traditional use of the technology. You will see in some of the case histories, that companies have opted out for the more open middleware solution. If they outgrow their switch, they can move to another manufacturer or switch, and they do not have to change software and retrain people to make the system work for them.

Computer Telephony Integration Workgroups

You may think that all this technology is great, but it sounds far too expensive and too tough to handle for a smaller business or smaller workgroup within a corporation, such as a technology help desk. The technology infrastructure is networked computers, and a computer acting as a server connecting these systems together. You may have only 5 people in the sales department, or customer service department. How can you use this technology and still afford it and show an good return on investment? The concepts of workgroup computer telephony integration are centered over the area of networked computer, and making the private branch exchange a network device to provide the call handling and routing. The key issues in this environment is providing the control software between the workstations and the private branch exchange.

There are 3 standards that have been used in the market, each vying for domination. The first standard is called Telephony Server Application Programming Interface (TSAPI), and was developed by AT&T. This standard was later implemented for commercial use by Novell server, who developed software drivers to allow workgroups to do screen pops, working with AT&T private branch exchanges. This has met with moderate success, and since Novell sold its UNIX rights to the SCO in 1995, this has not been ported by Novell to UNIX. However, it is possible that your UNIX system may be the database host in your network, and that you have a Novell server that can talk to the UNIX systems. This could allow TSAPI to be used on the Novell, to access both the database information, as well as the interface to the switch for screen pops. The factor with this technology is that it is computer server centric. The Novell server talks to the private branch exchange and then directs the calls as much as would a larger call center.

Another standard is TAPI, which is Microsoft's foray into the market. The difference here is the technology is desktop oriented instead of sever oriented. (Although recent emerging changes to the standard will see server based configurations emerging with Windows NT.) The desktop talks directly to the switch, using software loaded on the personal computer. Generally, this technology has not been adopted in a UNIX environment, other than like the Novell solution. The UNIX system hosts corporate data that is accessed by the client personal computer for screen pop capabilities.

Then there is another technology, which is used by Dialogic's CT-Connect middleware which supports Computer Supported Telephony Applications (CSTA) standard interface, for those switches that support this interface, as well as TSAPI, and TAPI technologies. The SCO and Dialogic announced on October 8, 1996, that they are working jointly to port CT-Connect to SCO UnixWare, and the product will be ready for shipment in 1997. This technology will allow UNIX developers to write applications with a middleware product, that scales from the small networked workgroup to the large call center solutions.

[8] Quote taken from The Economist article in the Business section, "The war of the wires: Two years ago America's local telephone companies dismissed the Internet as a fad. Now they worry it will ruin their business" in Issue: May 11, 1996: as http://www.economist.com/cgi-bin/lookup.cgi as posted on the world wide web at 09/23/96 12:05:35 PM EDT.

[9] Statements made by Harry Newton as witnessed by the author at SCO Forum95, during his Keynote daily speech on Thursday, August 17 1996, at the Quarry amphitheater at the University of California, Santa Cruz. Also can be seen in print in Computer Telephony '96 Conference & Exposition, The Show Directory Guide, "Why Pigs?" sidebar on Page 46, by Harry Newton, publisher. Published by Flatiron Publishing Inc. 12 West 21 Street, New York, New York 10010.

[10] "Gray, Elisha" entry in Microsoft Bookshelf '95 (Win32)- Encyclopedia, Compact Disk for Windows 95 Copyright Microsoft 1993-1995.

[11] Uniforum's IT Solutions June 1996 article "Sink or Swim with CTI" (253K): "When enterprise planning calls for computer/telephone integration, IS professionals must rise to the challenge." by Howard Baldwin http://www.uniforum.org/news/html/publications/ufm/jun96/start.html as posted on the Uniforum World Wide Web Home page of 09/11/96 3:42:57 PM EDT

[12] Client Server Computer Telephony: The Definitive Roadmap to the Client-Server Revolution, by Edwin Marguiles, page B-91, Copyright 1994 Edwin K. Marguiles, published by Flatiron Publishing Inc. 12 West 21 Street, New York, New York 10010, ISBN# 0936648-55-4.

[13] Harry Newton's Telecom Dictionary, page 452, by Harry Newton, Copyright 1994, Harry Netwon, published by Flatiron Publishing Inc. 12 West 21 Street, New York, New York 10010 ISBN# 0-936648-87-2

[14] Definitions and information on this portion on bus came from Appendix B of Client-Server Computer Telephony: The Definitive Roadmap to the Client-Server Revolution, by Edwin Marguiiles, Copyright 1994, Edwin K. Marguiles, published by Flatiron Publishing Inc. 12 West 21 Street, New York, New York 10010 ISBN# 0-936648-55-4

[15] BBN Major Innovations as posted on their Word Wide Web Page on the Internet http://www.bbn.com/bbn_hark/harkinnov.html Copyright 1996 BBN Systems Corporation, all rights reserved. Used with permission of BBN. BBN Systems 70 Fawcett Street, Cambridge, Massachusetts 02138.

[16] BBN Key Technology Milestones as posted on their Word Wide Web Page on the Internet http://www.bbn.com/bbn_hark/harkkey.html. Copyright 1996 BBN Systems Corporation, all rights reserved. Used with permission of BBN. BBN Systems 70 Fawcett Street, Cambridge, Massachusetts 02138.

[17] A Brief History of Time: from Big Bang to Black Holes, by Stephen W. Hawking, pages vii-viii, Copyright 1988- 1990, Bantam Books, 1540 Broadway, New York, New York 10036. ISBN 0-553-34614-8

Chapter 2

Interactive Voice Response

Interactive Voice Response is one of many the types or classes of computer telephony applications, and possibly the one you may have had some experience, if you have ever used a "banking by phone" system. Interactive voice response (**IVR**) enables a caller to enter information, using a Touch-Tone[18] phone keypad as a entry device such as account and password information, in response to voice prompts. In some applications spoken single words are used instead of the keypad entry using speaker independent recognition. A caller would say "nine", instead of pressing 9 on the touch-tone phone. The most common usage for speaker independent recognition is to support rotary phones which can not supply the touch tone signal. As you will see in the Parlance Corporation case history in this chapter, this has been taken even further with full sentence recognition, where the caller has a *dialogue* with the computer system.

---

[18] Touch-Tone™ is a Trade Mark of AT&T Corporation. For readability, the remaining references to it will simply 'touch tone'

The "Cobbler's Child"[19] - Remedy Corporation

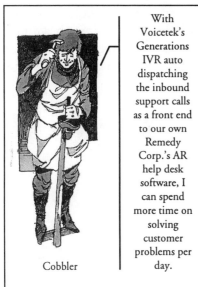

Cobbler

With Voicetek's Generations IVR auto dispatching the inbound support calls as a front end to our own Remedy Corp.'s AR help desk software, I can spend more time on solving customer problems per day.

Figure 2-1 Voicetek remedies the 'Cobbler's Child' syndrome for Remedy Corporation AR system.

You may know the parable that describes a poor cobbler who only has enough money to buy leather to make the next pair of shoes for a single customer. He never has enough leather to make a pair of shoes for his own child. This is a case history that provides an insight into how an interactive voice response application overcame this problem, by automating the help desk dispatch functionality, for a successful developer of help desk software.

BACKGROUND

In this case our 'cobbler' is a leading client/server help desk software development company, called Remedy Corporation of Mountain View, California. The 'cobbler's child' in this case was their own Action Request System software. Action Request (AR System) is a powerful, flexible, and easy-to-use client/server system, designed to manage the timely resolution of problems and support requests. They sell, and of course use this software internally to help support their customers, and already had achieved a lot of success for support of their customers. At their Mountain View, California headquarters, Remedy Corporation has a wall that is covered with praising letters from customers, of which one says, "'the best customer service organization"[20]. The reason for this is their ability to direct a customer to the proper support resource;

whether the support resource be a computer based system, or a help desk support engineer, in the least amount of time. Although Remedy was successfully handling the current case load, there was room for some improvements to increase the number of calls per day each support engineer could handle. In order to achieve that goal, the support engineers needed to spend more time solving customer problems than managing the administration tasks. These administrative tasks included:

- Identifying the caller and making sure that they are under support contract.

- Identifying which support engineers are qualified to handle the call. This will shorten the list of qualified support engineers.

- From the short list, find out what these engineers are currently working on, and what is already queued for them to work on in the next few hours. If the caller's contract specified that they must get a call back within 4 hours during business hours, then the person dispatching the call needs to assign the call to an engineer, who will be able to respond to the customer within that time frame.

Using the following Voicetek Generations IVR solution, Remedy streamlined this dispatch process by automating this operation to allow customers to help themselves.

OVERVIEW OF REMEDY'S ACTION REQUEST SYSTEM™

Remedy Corporation's Action Request (AR) system provides help desk and trouble ticketing capabilities for corporations. This product goes beyond these basic functions. At the heart of the system is an extensive and open database that provides the ability to allow corporate users to review the experience database for similar problems, on their own, before calling the organizations help desk.

As well, the AR system will maintain contracts that identify who is allowed to call for help and for what products. As calls for assistance arrive at the help organization within the company, the AR system will be used to document what the issue is about, and assign the appropriate customer support person to help solve the request. If the request is not responded to within a specified time window, then the request is escalated to a higher level of request of service, and if necessary routed to another support or managerial person within the organization. As new issues are defined and resolved, the history within the database allows companies to assign specific measures, or "metrics", to identify specific issues so as to improve the overall quality of a product.

So, if a company was launching a new software product, and management had some suspicions of where they may expect some typical problems with installation of the product, they may want to set up some type of indicators within the database to be flagged when problems occur. As well, unforeseen incidents can be tracked, measured, and categorized to help development and quality assurance teams understand where product improvements must be made. Then, historical reports can be generated about the product's quality on these different indicators to show trends, or to help manage potential future problems.

"'The Action Request System is both an application and development tool'", and according to Peter White, Remedy's technical support engineer, 'can do anything' - from sales tracking and work flow management, to management of a wide variety of business processes, including technical support, defect tracking, change tracking and others. Special features include e-mail and pager alerts, which advise the user of critical calls if they have not checked their messages for a pre-set amount of time, as well as 'macros' that support repetitive tasks.'"[21]

THE INTERSECTION OF COMPUTER AND TELEPHONY - DATA VALIDATION

This case history demonstrates how it is easy to add a lot of value to an existing system by using interactive voice response. One of the most important and time consuming tasks at hand is the data validation process, to make certain that the support engineer is supporting a customer who is entitled to it. Peter White notes two easily identifiable benefits of the integration between the telephone and the Remedy Action Request's database. "'The joining of Action Request System and Generations represents a perfect integration of products', White says, noting that it 'enables us to do much more than just routing calls'. He explains that Generations serves as 'an excellent front end to both the Action Request System and our Rolm Automatic Call Distribution (ACD) system, and provides us with the key to increased port productivity'".

"Noting that 'data validation marks the real point of intersection between the two systems,' the Remedy manager explains that the interactive voice response (IVR) system helps Remedy technical support representatives immediately identify callers, obtain detailed information about them, and therefore assist each other in the most effective and efficient management possible' "[22]

How this is done, is described in Figure 2-2 on page 81. The telephony and data intersection, show how the calls are routed and then handled. More importantly, the caller is identified against the Remedy database, and if the person seeking help does not have a contract, then they are routed to a person within the company to purchase a support contract. This avoids having a person who is not entitled to service from taking valuable time from the support of others who have paid for it. As well, the system routes the caller to the fax-on-demand system, so as to take advantage of that service if he or she wishes to use it. Otherwise the caller is routed to a support engineer who has experience within the problem category as selected by the caller. The benefits are:

1. Faster and accurate transfer of callers to the correct engineers, with the most current information about an open job ticket, reducing the amount of telephone tag.

2. Callers get the appropriately skilled engineer in the least amount of time, because the callers select the problem category.

3. Using the Fax-on-demand service, customers can order a written solution to a comparable previously solved problem, saving time for the customers and the Remedy support team.

4. Support staff work with paying customers as validated by the system, and those who do not have contracts are asked to

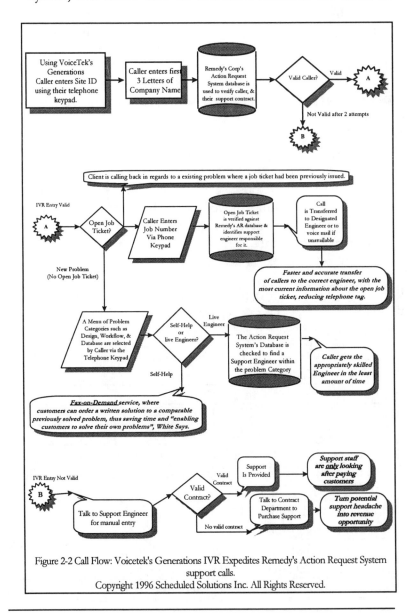

Figure 2-2 Call Flow: Voicetek's Generations IVR Expedites Remedy's Action Request System support calls.
Copyright 1996 Scheduled Solutions Inc. All Rights Reserved.

purchase a contract. This means a support problem becomes a revenue generating result.

CALL FLOW

Please find the call flow of how Voicetek's Generations IVR system handled the Call Flow to help customers to help themselves, using the interactive voice response system and fax-on-demand system to increase productivity of Remedy's support engineers.

THE BUSINESS REASONS WHY THE COBBLER IS NOT SO POOR ANYMORE!

In the brave new world of new technology, sometimes the excitement and elegance of the technology itself and the belief that it will actually have a positive effect, is sometimes very hard to quantify in terms of dollars saved and dollars earned. With this case history, the facts and figures are quantifiable in terms of manpower saved because of Generations™ ability to route the calls, and as well, the use of the fax-on-demand system.

REMEDY CORPORATION CALL VOLUMES/HISTORY	
Number of calls per week	750
Number of calls per Month (1 Month is averaged out as 4.2 weeks/month)	3150
Number of calls per year	37800
Number of Engineers in July 1995	28
Number of Engineers in June 1996	33
Percentage of calls needing Engineering contact	70%
Percentage of calls who could be referred to the FOD after talking to an engineer.	30%
Monthly total Number of Calls directly answered by an Engineer (June 1996)	2205
Monthly Number of calls referred to self-help within the FOD after talking to an Engineer.	661.5
Monthly Total number of Calls who required help	1543.5

REMEDY CORPORATION CALL VOLUMES/HISTORY	
from an Engineer	
Average duration of telephone time for a support call (approximately) expressed fractions of an hour	0.75
Estimated cost of a direct support call - averaged estimate in US Dollars.	$120.00
Monthly Support Call Self-help cost savings on Engineer referrals to FOD as expressed as a number of such referrals times the cost of the average cost per call.	$79,380.00
Monthly Support Call Self-help hour savings on Engineer referrals to FOD as expressed as a number of such referrals times the average length of a call	496.125
Because of the self-help referral to FOD, each Engineer does not have to handle this number of calls per month	20.05
Because of the self-help referral to FOD, this is the amount of labor saved each year expressed in man months	2.953125
FOD self help provides the labor of these many Engineering man months each year	35.4375
Hence the IVR self-help system has reduced the need to hire another this number Engineers in the first year of operation.	2.95
Because of the self-help referral to FOD, on average each Engineer get these hours per month to spend on providing higher quality of service to the customers that need it most. This time would be used to talk to customers or to research the problem.	15.03
This is the number of calls per month that the IVR system handles without engineers.	945

Table 2-1 "Cobbler Child": Remedy Corporation's Call History Statistics.

In preparation of this case history, Peter White was able to look back on almost one year of operation of the system. The old saying that hindsight is 20/20 applies, and in this case shows a very valuable and profitable lesson. The lesson can be described in quantitative measures, as to how the fax-on-demand saved engineering support time, while simultaneously increasing the value of the quality of service provided. For a complete dissection of the capability of the product of their efforts please refer to Table 2-1 "Cobbler Child": Remedy Corporation's Call History Statistics.on pages 83.

With all the statistics above, one may find it confusing to understand the true value of Voicetek's Generations™ system, as it impacts Remedy's support services. These are some of the highlights of the statistics for those calls that actually got to a support engineer:

I.   After a short consultation, 30% of the calls were referred to the self-help Fax-on-Demand services. With reported level of calls per month, this equals the net equivalent of 3 virtual support engineers.

When one uses the median salary, (50% below and 50% above a salary survey sample), in the San Francisco and San Jose area for pre/post sales support representative, a support engineer's salary is about $80,000 per year[23]. In manpower alone the saving to Remedy is about $240,000, based on this salary survey It is important to emphasize that we are using a salary survey, as referred to in the endnotes, as a basis of this saving, since we are not privy to the actual remuneration of the skilled labor at Remedy. The San Francisco and San Jose area computer professional labor prices are typically the highest, compared with most other places in the United States. At the other end of the price spectrum, the same position in the Memphis and Nashville Tennessee area would garner a median salary of $41,000, indicating the regionalism of the labor market. The survey indicates the most common salary is $52,000 to $63,000 in other centers of the United States.

Moreover, the savings of this manpower and growth as reported in Remedy's Annual Report dated December 31st, 1995, shows a company with an explosive growth of 103% over the previous year. The financial statements show the reported net income for that year was $7,561,000[24]. If one assumes that the salary survey is correct in terms of labor costs, then the last 5 months of operation of the system in 1995 represented an averaged saving of $100,000 during that period. This represents 1.3% of the reported net profit for the operational period of the system. If one extrapolates the labor saving during the period of 1 year of operation, then the system saved about 3.17% of net profit, given that the net profit of the company was the same as the last fiscal year. Furthermore, Peter White also indicated that besides the Fax-On-Demand Self-help services, their Internet World Wide Web site has had about 4 times the number of documents reviewed per month. In theory, that too has mitigated the need to hire more Support Engineers by at least 9 more staff members. Peter sees the Fax-on-Demand and World Wide Web self-help services as being complementary, rather than being in competition with each other. The convenience of using the telephone, and not having to be skilled in connecting to the Internet is why the Fax-on-Demand service is so valuable.

II.   The Fax-on-Demand self-help service provides a true savings, in terms of time, to each engineer in the support team by giving him, on average, 15 more hours per month to deal with customer problems. This provides a higher quality of service to these customers and allows more time to research solutions. As well, the number of calls is also increased to 20 per month per engineer.

One may argue that the 662 calls per month that are referred to the self-help fax-on-demand system, could be a lost revenue of $79,380 per month. Why not just hire more support engineers, and charge customers the support fees to earn the income, which would surely cover the salaries of 3 extra support staff? Although the argument may have some merit, this is really not lost revenue. First off all,

Remedy has increased their support staff from 28 to 33 members in the team, to meet demand for such services, because of true growth of their business. As well, each of those engineers has benefited from the presence of the solution presented here. One may ask these rhetorical questions:

1. Of those 662 calls, how many customers were actually under contract? The question is simple to answer. In order to access the self-help fax-on-demand system, callers had to have a valid support contract to access this important service. Therefore, there is no true lost revenue. Instead, this is an argument for more profitability.

How many would be willing to pay the cost of a support engineer, when the information could be quickly and easily obtainable at a lower cost? Some may argue that no matter how easy the problem is to solve, a caller will want support from a person, rather than a computer system. This may be true for some, but not all. As the utilization statistic shows, 30% call referrals to the fax-on-demand self-help system, has allowed many customers to obtain self-help information. As shown by the growth in the company, the demand on the interactive voice response fax-on-demand system has had over 100% growth in 1 year, from 300 per month as reported in July, 1995 to 662 in June, 1996. In order to access the self-help fax-on-demand service, you have to be under contract to use that portion of the system. Moreover, since the technology described here is not exclusive to Remedy, and their competition also has access to it from other computer telephony vendors, then it is obvious, that this $79,380 per month, is money saved by Remedy's customers, which they use to make themselves more efficient. Hence there is a *ripple effect* for not only Remedy, but to their customers as well.

It is better to express this $79,380 per month as money saved by Remedy's customers to make them more competitive. As with computer telephony application deployment, the vendor who sells a client on such technology, is ultimately not only making its direct customer derive cost benefits, but as well, it saves or improves the services of the client's customers. Given that *ripple effect* to

Remedy's customers for a single year is $952,560, then that is no cheap shoe leather!

Moreover, the net effect of not having to hire new engineers, means that Remedy has not expended about $240,000 per year at the current call levels.

Given that the system was installed within 1 week, the return on investment has been very significant. The cobbler has more income and more happy customers than before. $240,000 in estimated savings can be used to purchase some pretty nifty shoes for the cobbler's children - Remedy's shareholders!

Vital Statistics

The "Cobbler's Child" - Remedy Corporation	
Cost	*Statistics*
Weeks of labor to install system	1
Initial System Cost	not reported
Est. Ongoing support Cost	not reported
System Size	
# Ports - Total/# Fax.	12 Ports - 2 of which are for fax
# Users	not reported
# Calls/month	3150
Computer	Intel 486/Pentium PC
Operating System	SCO UNIX 3.2.x for production system, and Sun based System for Development.
Industry	*Software - Help Desk*
Estimated Return on Investment	1.3% of 1995 Net Profit
Cost Savings	*not reported*
Manpower equivalency per year	3
Estimated Manpower savings per year	$240,000
Reduction in telephone costs/year	not reported
Earnings	
Generated new income/year	not reported
Ripple Effect	
Estimated cost Saving to customer base	$952,560
Vendor	*Voicetek*
Product	Generations
End User	Remedy Corporation

The "Cobbler's Child" - Remedy Corporation

Cost	Statistics
Computer Telephony Technologies Deployed	
Interactive Voice Response	Yes
Fax On Demand	Yes

Table 2-2 The "Cobbler's Child" - Remedy Corporation - Vital Statistics

---

[19] "The Cobbler's Child" concept came originally from Catherine Jaspersohn formerly of Voicetek, and now with Vivant Associates. Email: cjaspers@shore.net. I owe this title and idea to her. The work I have contributed to hers are the detailed functionality, accounting and ROI.

[20] "Remedy - Voicetek Corporation Customer Case History" page 1. Copyright 1996 Voicetek Corporation, 19 Alpha Road, Chelmsford, MA 01824.

[21] "Remedy - Voicetek Corporation Customer Case History" page 1. Copyright 1996 Voicetek Corporation, 19 Alpha Road, Chelmsford, MA 01824.

[22] "Remedy - Voicetek Corporation Customer Case History" pages 1-2. Copyright 1996 Voicetek Corporation, 19 Alpha Road, Chelmsford, MA 01824.

[23] 1996 Computer Salary Survey and Career Planning Guide, Source EDP® Computer Recruiting Specialists, A Division of Source Services Corporation®, head office, 5580 LBJ Freeway, Suite 300, Dallas, Texas, 75240, 214-395-3002, http://www.sourcesvc.com. Please note this salary survey has been a reliable source in the computer industry for many years. The cost of the pre/post support sales professional in this part of the USA as reported by the Survey, indicates that the San Francisco/San Jose area has the highest median salary for this position in the USA.

[24] Remedy Corporation Annual Report Financial Highlights, as posted on their World Wide Web page on the Internet http://web.Remedy.COM/MARKETING/ann_report/highlights.html

The "Natural Advantage" - Parlance Corporation's Name Connector

In this case history we will review and explore the reasons why some interactive voice response technology functions do not meet everyone's needs. There are some situations where more advanced technologies called speaker independent recognition (SIR), and Text-to-Speech (TTS), work together to create an environment that is much easier for people, rather than for the

With Parlance Corp.'s Name Connector, all I have to say is "Dr. Sieniewicz" and I'm immediately connected to him. No more remembering endless lists of telephone extensions!

Figure 2-3 Speaker Independent Recognition: Parlance Corp. provides a more natural way of working with an Interactive Voice Response application.

technology. Speaker independent recognition allows spoken commands to be interpreted by the computer and acted upon. You should not have grand visions of speaking to a computer in the same manner as the fictional interstellar crew of the television and movie series *Star Trek*. Rather, one should look at a simple function that we all face in everyday business. In order to demonstrate what that is, we will first go through an imaginary scenario, (but all too real), to show where and why interactive voice response technology has its limitations. Then, we will focus on Parlance Corporation's Name Connector and how at the offices of Bolt Beranek & Newman Inc (BBN) they have been able to use this technology for 3 years to solve a problem, which to some may seem rather small, but in reality has a dramatic cost saving and psychological benefit to all their employees. Parlance Corporation is a wholly owned subsidiary of BBN.

Congratulations Doctor - You saved the patient and discovered a new disease called IVRitis!!

Imagine you are physician specializing in oncology, (cancer and tumors), and you need to contact an internationally noted physician and radiologist, (X-ray & imaging), for an important consultation. It is midnight at the radiologist's office at a large university in Montreal, Canada, which is located 6 time zones behind your local time in Paris, France. He just called you, and said that he needed to review some new test results just delivered to him. He asked you to call him back in about 30 minutes.

Time is up, and you place the call. The university you are calling has voice mail. You are asked to enter the person's extension. You do not know it. You press 'zero' to bail out and try to get a human to answer your desperate needs, and no one answers. Frustratingly, you are greeted with a general voice mail box greeting, saying something to the effect that it is after hours, and no one is around to help you. You must record a message in the black hole of the 'General Delivery Mailbox'. You think to yourself, 'If the receptionist remembers to clear this message sometime *tomorrow....*' You mutter something about 'Voice Mail Jail' as you hang up half-way through its attempt to record your message.

You call back, and carefully listen to the menu choices, and you press '#' on your telephone keypad to access a company directory. *You think you are finally saved from this nightmare - but it has only just begun!* You only know the last name as you remembered it when the radiologist introduced himself on the phone, now 40 minutes ago. It was a phonetically difficult name. You are asked to enter the first 3 letters of the last name by the voice mail's interactive voice response system, using the telephone key pad's alphabetic characters. You think to yourself, "Now is it 'Synavich'? Or should I spell it 'Sinawich'?. Or should I try 'Cynavich'". Remember, you are calling from Paris, France, where the telephones do not have the alphabetic characters on the telephone key pad.

Since you received your education and did your internship at Harvard Medical School in Cambridge, MA., you vaguely remember how the telephones back in the USA had the alphabetic characters mapped on the keypad. Valiantly, you try to spell 'Syn' on the keypad of the telephone using the interactive voice response's character identification of typing in Morse code fashion the *letters* on the *numeric* telephone keypad. The voice mail system reports that there is no such person, and you try again with 'Sin', and again it fails. You've got to make contact! Otherwise your patient is going to get much worse very soon. You can not wait until the next day. So you think to yourself, "Maybe if I only put in the first letter-like the letter 'S', perhaps the system will tell me all the names and extensions starting with that letter." You give it a try, and you rue the day that you did! The letter 'S' starts more words and names than any other letter in the English and French languages. You listen to over 50 names and extensions, as the interactive voice response software plows through all the names in the following fashion: "For 'Dr. Susan Sampson' please enter extension 7726". Finally after 2 minutes and 30 seconds of this litany of names and extensions you are shocked to hear the doctor's own voice played back, "For *Dr. Sieniewicz*", please enter extension *7744*".You enter the extension number and finally talk to the radiologist. A potential cure for the mother of a family of four young children is discussed and adopted. What would have happened to that family, if the technology was not conquered by your extreme determination and motivation not to be defeated by this wonder of technology! *Mon Dieu!* There has to be a better way!

There is a better way, and it is called Speaker Independent Recognition. Using a product like Parlance Corporation's Name Connector, you would have only needed to call this imaginary university, and said the phonetically equivalent of "Sieniewicz" and would have been immediately connected to this important radiologist. (As you lived in the United States for 9 years , you only

➢ Travel Alert!

In reality, telephones in Paris do not have alphabetic characters mapped to the telephone keypad. Try calling your office back in North America, from Charles de Gaulle airport using a interactive voice response directory. It truly is a frustrating experience after a transcontinental night flight!

have a very faint trace of French accent when you speak English. This does not confuse the speaker independent recognition software.)

The Cure for IVRitis - Parlance Corporation's Name Connector

What made the poor Parisian oncologist go completely crazy? The poor fellow was forced to fit the rules of the computer software, rather than the computer software fitting the human. As you can see, because of the limitation of the telephone keypad's ability to interpret letters, (*even if they are marked on the telephone keypad*), this does not lend itself to a simple interface. Moreover, if the doctor knew the extension number, he is still forced to remember that, rather than accessing in a much easier way. What would be much easier? The answer is large vocabulary continuous speaker independent recognition. If our Parisian oncologist called the Montreal radiologist's offices at the university, and if the university had Name Connector, then the oncologist would only need to say the name of Dr. Sieniewicz phonetically as 'Doctor Sinavich' when prompted, and he would have been immediately connected.

Call Flow

> *Historical Trivia*
>
> BBN is one of the original research and development forgers of the technology

The Parlance Corporation Name Connector does just as it name implies - it routes calls. Parlance Corporation developed the Name Connector as an internal project to speed the handling of the call direction within the whole BBN corporation. The Name Connector is accessible to employees within BBN's offices, and accessible to them from outside the office via a local call or by a toll free 1-800 call. It is not a voice mail system, nor as Drew Knowland, former Marketing Manager of Parlance says, "It is not an interactive voice response application" in the traditional sense of the term, indicating that the Name Connector is much more sophisticated technology, and does not have all the drawbacks and the restrictions of a traditional interactive voice response system. As well, their employees call the system, and when prompted only have to say the name of the party to whom they want to be connected. The name of the party is found in the database, and then spoken back using Digital Equipment Corporation's

DECTalk™ text-to-speech for confirmation and the call is transferred to the party.

As a company, BBN has many offices throughout the United States of America, with different operating units such as BBN Planet, specializing in Internet and has other divisions. An employee can call any of the other 2000 employees within the BBN Corporation, and have the call directed to any phone within BBN's Cambridge, Massachusetts head office, or to other offices throughout the USA and internationally. Please see Figure 2-4 below for details. This provides the same functionality as a live operator, with the same interface, without having to pay for staff to answer the calls. It takes this level of technology in order to deliver the functionality that is required to provide the easiest form of interface, and deliver the promise that interactive voice response systems can not completely fulfill.

The caller only hears a brief prompt to speak the name of the individual he or she is trying to reach. If one asked for "Drew Knowland," the Name Connector searches the database for a match, finds it and then speaks back "{transferring to Drew Knowland}" using text-to-speech, and the caller is connected. If

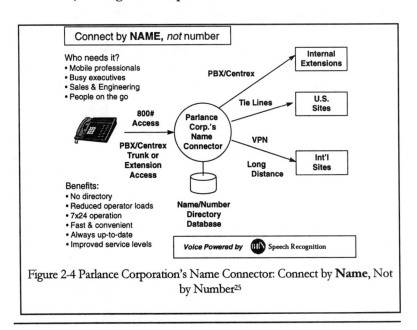

Figure 2-4 Parlance Corporation's Name Connector: Connect by **Name**, Not by Number[25]

the caller stumbled and was unsure who he wanted to call by saying "Uh, umm, ahh, Oh!," with a long pause as he fumbled through papers looking for the name, then Name Connector will look in the database for a person's name or nickname that matches "Uh, umm, ahh, Oh!". It will try to make a match with the closest sounding name or nickname and if there is a match, it will connect you to that person's extension. Name Connector will give you a chance to cancel the selection by pressing "*" on the telephone keypad, and transfer you to its next best choice. You can cancel the choice and be prompted to say "Help" to get some idea of how to use the system, or "Operator" to transfer to a live operator.

The management of the name list for the database is a task that is performed nightly, as an upload from the Human Resources department database. Any changes to names and extensions are added to the system. The process is mostly automatic and generally no further editing needs to be done to the database data. As in nature, there is always an exception to the rule, and manual editing has to be performed by the system administrator once every week to make small changes . If Rob has a long and difficult phonetically sounding name, it may befuddle a human speaker's pronunciation. Adding a nickname such as "Rob Y" may be the answer. :As we all have experienced, indigenous English speakers may have difficulty with the pronunciation of words that sound different than they are spelled. This is a limitation of human kind, rather than of the technology. Currently, BBN makes about 10 to 20 such refinements every week, needing a couple of hours of the system administrator's time.

Because of the ease of use, former employees who have moved to new career opportunities at other companies are "Name Connector Addicts", says Drew Knowland. While at BBN they had become so accustomed to calling others in the company by name, they really yearn for it at their new companies, since they are now forced to use a standard interactive voice response autoattendent system to navigate through their organization.

The Business Reasons for Using Name Connector

Why should a business fund their "addicts" and their "addiction"? The business side of you must be wanting to know why would anyone spend all this money, time and research so one can make one's life a little easier. Like any purveyor of an addictive substance they will do it for the shear joy of the profit of it, and to subsidize their own habit. In this case the 'narcotic' is not self destructive and completely socially and economically acceptable.

RETURN ON INVESTMENT

Where is the profit to be made? It is all in the numbers.

The Parlance Name Connector resides on a two Silicon Graphics Inc. Indigo 4000 system with 6 analog telephone lines adapters built into the system. They only need to install the Name Connector application software and database. These systems answer 3900 calls per day! This provides a level of service to the company staff that off loads the labor intensive task in a more natural way of working with the system, while at the same time freeing up the labor needed to attend to outside companies, or special requests that require special attention. It is the special service which enhances the company image.

Since around 25% to 50% of the calls into BBN are performed by BBN staff members, it makes incredible sense to use the Name Connector to route the calls on low cost internal leased lines. This utilizes a least cost routing mechanism to further drive down costs. With most of these calls being made on a toll free 800 line, this further restricts the cost curve, and reduces the chance for calling card fraud and the management issues that also entails. Jack Reilly, President of Parlance, stated this was the largest cost saving feature of using a system such as Name Connector.

Why have a paper hard copy of a telephone directory of names and extensions when you only need to say the name of the person you want to call? The purpose of having a directory is find an individual or group of people. It is not likely that a single staff member will need to call every other staff member in the whole

BBN organization. This implies that a lot of names in a printed directory are redundant and possibly not relevant to a single staff member's needs. Staff tend to work in groups, and to have a circle of associates that work together on projects. Generally, they know who they want to talk to most of the time. Since the system is updated every night with new names, then that new employee is instantly made part of the telephone access. The need to publish a hard copy of the company telephone directory was cut in half. Like all paper copies of telephone directories, they are out of date by the time you have printed and distributed them. Since there are 2,000 employees in BBN, this means that a directory the size of about 120 pages had to be distributed to all 2,000 employees throughout the USA. Drew Knowland estimates that this reduction in cost to the company was about $24,000 per year, in publishing and distribution costs.

RIPPLE EFFECT

The Parlance Name Connector routes more calls than a small team of operators could ever do, and even faster than a standard interactive voice response system. On average it takes about 10 to 13 seconds, (11.5 on average), for the Name Connector to route a call. The time measured here is the interval between the time the call is answered and the time the call is transferred, and the leading sound of the ringing of the extension. What transpires is Name Connector picks up the call and says "{Name Connector, Name Please}" -Tone- and Name Connector listens for your request and start dialing the extension. (If the party you wanted to speak to was not there, the call is routed to a voice mail box.). A standard interactive voice response application, used in an autoattendant, will take anywhere from 15 to 20 seconds to play the instructions, accept touch-tone input, and instruct the private branch exchange to transfer the call to the intended extension. Even if you know the extension and the pattern of touch-tone entry, and you immediately enter the extension you want without waiting for the instructions to play, then it would be about 10 to 15 seconds for the call to be routed to the extension you wanted. Shaving 5 seconds does not sound like much but think of the call volume, and multiply it by the number of seconds to perform a call transfer. That represents about 5.42 hours each day of dead wait time on the phone that the organization is incurring every single business day of the year. This represents about 169 mandays per year, or the equivalent of 68% of one persons time in a 50 week year.

It is these facts that show the implications of how the next generation of technology will definitely provide new levels of capability and cost savings. Although this interface can not really reason like a human, at least the technology has the limited ability of human capacity to listen and speak the way we do. This is a blessing which we will soon be able to manage at lower cost, and with greater functionality. This technology can cure "IVRitis". The pleasant side effect is that it could make happy, socially adjusted and productive "addicts" of us all, from which I hope there is no cure!

Vital Statistics

The "Natural Advantage" - Parlance Corporation's Name Connector

Cost	Statistics
Weeks of labor to install system	not reported
Initial System Cost	not reported
Est. Ongoing support Cost	not reported
System Size	
# Ports - Total/# Fax.	6
# Users	2,000
# Calls/month	85,800
Computer	Silicon Graphics Indigo 4000, MIPS 4000
Operating System	IRIX 5.2, a UNIX System V Release 4.x
Industry	*Software*
Estimated Return on Investment	not reported
Cost Savings	*$24,000 per year in directory publishing costs*
Manpower equivalency per year	not reported
Estimated Manpower savings per year	not reported

The "Natural Advantage" - Parlance Corporation's Name Connector

Cost	Statistics
Reduction in telephone costs/year	not reported
Earnings	
Generated new income/year	not reported
Ripple Effect	
Estimated cost Saving to customer base	169 mandays per year
Vendor	*Parlance Corporation*
Product	Name Connector
End User	BBN Corporation
Computer Telephony Technologies Deployed	
Large Vocabulary Continuous Speaker Independent Recognition	Yes
Text-to-Speech (DECTalk™)	Yes

Table 2-3 The "Natural Advantage" - Parlance Corporation's Name Connector - Vital Statistics

---

[25] Connect by **Name, Not Number** - Parlance Corporation presentation slide Copyright 1996 Parlance Corporation, all rights reserved. Used with permission of Parlance. Parlance Corporation 70 Fawcett Street, Cambridge, Massachusetts 02138.

The "Orient Express" - Kowloon-Canton Railway Corporation

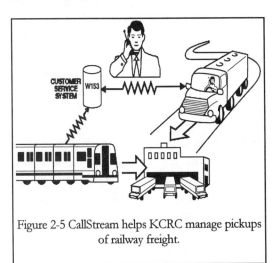

Figure 2-5 CallStream helps KCRC manage pickups of railway freight.

Imagine having 12 freight trains per day bringing about 150 railway cars of cargo arriving from the People's Republic of China (hereafter referred to as "the Mainland"). How do you handle the many inquires about the expected delivery of these goods? How do you maximize the delivery while minimizing traffic for picking up the cargoes at the freight terminal? Managing a railway is a complex business. If the railway can use any automation to help their customers obtain their cargo faster and easier, the less expensive it is for all involved. This case history shows how a UNIX computer telephony solutions do not have to be high density, large installations to make a significant impact. Rather a small number of interactive voice response ports, hosted with the main database application, is the type of solution that can be easily integrated, providing a significant payoff.

Background

The Kowloon-Canton Railway Corporation (KCRC) was started in 1911 to provide transportation between the Mainland and Hong Kong. Kowloon is a city of Hong Kong on the Kowloon Peninsula opposite the island of Hong Kong, located on the southeastern coast of the Mainland. The rail link to the Mainland was closed after the communist government took power in 1949. It was reestablished in 1960, and is now a vital link between manufacturing plants within the Mainland and the financial and shipping center of Hong Kong. "The KCRC in Hong Kong is a

> Historical Fact.
>
> Hong Kong was a British Colony which has reverted back to China rule and administration on July 1, 199⁷

99

quality pioneer with a firm commitment to customer satisfaction, efficient service and engineering excellence, with an ISO 9000 qualification. It provides passenger and freight rail transportation services, a light rail transit system and feeder bus services, handling over a million passengers per day. It is a public corporation which is run on commercial principles and is self-financing, the KCRC also engages in property development along the railway and has embarked upon a program to expand its capacity and upgrade its services."[26]

The rail traffic between the British Colony of Hong Kong and the Mainland is increasing. Because of the growth and competition from the trucking industry, there is a strong commitment by the railway to build new rail infrastructure to facilitate the expected continued growth, well into the next century. This is evident in the construction of the West Rail Corridor for rail containerization traffic to the Mainland. Once the new Kwai Chung Port Container Terminal is completed in 2002, this will be one of the most advanced rail containerization ports in the world. It will provide better service than transport trucks, since the containers will be able to be loaded directly from the railway cars to the ships for export. This eliminates the intermediate step of having to move cargo from the railway terminal to truck and then to the ships. They forecast the rail traffic will increase to 350 railway cars per day by the year 2002.

After products have been manufactured in the Mainland they leave for Hong Kong. At that time the shippers in the Mainland will notify their customers in Hong Kong which railway car or cars contain their goods. This is one source of key information the customers obtain to find out where are their cargoes on which railway car. Once the train has arrived at Shenzhen at the border, the manifest describing the contents of each railway car is entered into the KCRC's Wagon Information System III (WIS3). Then the railway's local agents can notify customers of the wagon number, the date and time of arrival. The Wagon Information System III is an on-line database system running on SCO UNIX. It was established in 1994, and this system laid the groundwork for the implementation of an interactive voice response system in 1995.

System Overview

With all the inbound freight traffic, a crucial issue was scheduling the pickup of cargo from the rail terminal. Before the interactive voice response system was installed, the railway's customers experienced extra costs due to delays in the collection of cargo. These costs involved having trucks, drivers and loaders wasting time before they could pick up their cargoes. Staff in the customer service department were taking over 100 calls per day to help customers identify customer pickup times. This distracted them from solving more pressing and urgent customer service problems.

The Kowloon-Canton Railway Corporation prides itself on excellent customer service. Percy Yeung, Freight Operations Officer (Planning), for Kowloon-Canton Railway, said they realized they could solve the problem by providing customers with timely information of a schedule to pick up their cargoes. Customers can make inquiries 24 hours a day 7 days a week, in order to obtain the timely information they needed. Percy Yeung said, "There were many interactive voice response systems in the market. Our goal was to adopt the one which could link our on-line database on SCO UNIX. Most of the interactive voice response systems could not economically synchronize the data with the SCO UNIX system. CallStream [CallStream's VoiceStream IVR] system solved our problem." Besides the ability to work natively on SCO UNIX, the system had to be customized to be bilingual in both Cantonese Chinese and English. The system was readily adaptable to this. Since VoiceStream IVR originates in Canada, a bilingual country, the French Canadian system pre-recorded concatenated speech phrases were replaced with Cantonese Chinese system phrases.

> ➤ *Usage Alert!* Railway cars are referred to as "wagons" in Hong Kong. I have taken the liberty to use the North American usage of "railway cars" in the rest of this case history.

Once the system had been installed, the Kowloon-Canton Railway Corporation allowed a selected set of customers to use the system for a trial period. After they felt the system had met their criteria, they did a marketing and advertising campaign to notify the rest of their customers about the system. There is always value in testing a prototype system with a limited set of customers, before rolling the system out to the whole customer base. Moreover, the issue of marketing the access phone number is important, which helps ensure utilization of the system.

The interactive voice response system now gives the customers up to 4 hours advance notice to pick up their cargo at the freight terminals. This alleviates the congestion at the freight terminals, and lowers costs for their customers.

Call Flow

The call flow is very easy and fast to use, which adds to its charm. The caller is asked to select Chinese or English. From a menu of

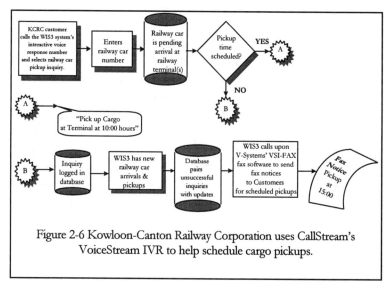

Figure 2-6 Kowloon-Canton Railway Corporation uses CallStream's VoiceStream IVR to help schedule cargo pickups.

choices, the customer can then select a railway car inquiry. They enter the railway car number, and the system tells them the time and date to pick up the cargo at a specified terminal. The notification of the pickup has about a 4 hour advance notice. If the customer makes an inquiry before the pickup time has been determined, the inquiry is logged in the database, and the customer is told that a fax will be sent to their destination fax number indicating when to pick up the cargo. Once the pickup time has been determined for such a logged inquiry, the database application calls upon V-Systems VSI-FAX to send the fax using the fax number in the database for the customer. Besides scheduling pickup information for the railway's customers, the system has an optional fax-on-demand system that allows customers to retrieve other general information about the railway and about the system itself. Of course the documentation has to be in both languages.

One of the advantages of keeping the touch-tone data entry relatively simple, is that it allows the customer's truck drivers to use their cellular phones to learn when to pick up their cargo, while they are doing other deliveries.

Business Analysis

The business goal here was too improve customer services, which is what the railway carefully measures. Providing such a service in a competitive market brings customer loyalty to the railway, and lowers everyone's costs.

RETURN ON INVESTMENT

Although Percy Yeung could not directly attribute a reduction in costs in manpower, it has been a very successful implementation. The system has about 100 inquiries per day. Over 95% of the call volume for information in regard to scheduled pickups are now being handled by the WIS3 interactive voice response system. The remaining percentages who still need to talk to a customer services representative are those few customers who share multiple consignments in the same railway car, and these are not resolved by the system at this time. This is a limitation of business requirements surrounding the mixed consignments, and not a limitation of the interactive voice response system or of the database software. Percy also noted that the growth of calls to the system are 5% per month, indicating the growth of the railway's business, and that the system is able to grow with the demands being placed upon it.

If, under the manual system, each inquiry took 3 minutes to answer each of 100 inquiries, then this would represent 5 man hours in every 24 hour period This time saving has been deployed to look after other customer problems that need more personal attention.

RIPPLE EFFECT

The response to the WIS3 interactive voice response system by the customers has been very good. With 95% of inquiries being handled by an interactive voice response system, it demonstrates the system's ability to quickly and efficiently handle customers' requirements. As Percy Yeung said, "According to our annual Customer Satisfaction Index (CSI), the interactive voice response system showed a significant return from our customer's satisfaction". Beyond the subjective expression of the customers in the survey, the customers saved money in terms of time spent calling the railway's customer services, and more importantly saved time in appropriately scheduling cargo pickups. This resulted in better utilization of manpower and equipment to help them serve their customers.

Vital Statistics

The "Orient Express" - Kowloon-Canton Railway Corporation	
Cost	*Statistics*
Weeks of labor to install system	4
Initial System Cost	not reported
Est. Ongoing support Cost	not reported
System Size	
# Ports - Total/# Fax.	4 ports voice & Fax
# Users	Hundreds
# Calls/month	3,000
Computer	Pentium PC
Operating System	SCO OpenServer 5.0
Industry	*Transportation*
Estimated Return on Investment	
Cost Savings	
Manpower equivalency per year	5 man hours per 24 hours
Estimated Manpower savings per year	not estimated
Reduction in telephone costs/year	not reported
Earnings	
Generated new income/year	not reported
Ripple Effect	

The "Orient Express" - Kowloon-Canton Railway Corporation

Cost	Statistics
Estimated cost Saving to customer base	Savings in Manpower & Equipment
Vendor	**CallStream**
Product	VoiceStream IVR
End User	Kowloon Canton Railway Corporation
Computer Telephony Technologies Deployed	
Interactive Voice Response	Yes
Fax Server	Yes
Fax-on-Demand	Yes

Table 2-4 The "Orient Express" - Kowloon-Canton Railway Corporation - Vital Statistics

---

[26] Quote taken from *The Economist*, September 21st 1996 issue "Leadership by Stealth" Executive Focus, page 10, K/F Selection advertisement for "KCRC".

The "Registrar" - York University

Did you go to university or college before the mid 1980's? If you did, you will remember the new experiences that were a lot of fun, while some of them were not so great. One of those experiences that was not so great, was course registration. This case history reviews a very common and typical early adopter of interactive voice response technology. For those who were fortunate at their university or community college to use this kind of system, they may not know what the experiences were like in the days when this technology was not available. For their entertainment we review the "good old days" of course registration.

What makes this case history at York University so interesting is that the interactive voice response Student Registration system is now on its 3rd generation of technology, using an RS/6000 AIX based UNIX system, with client-server technology. This provides an interesting perspective in terms of the technology evolution, the system management and resources requirements.

Background

If you attended a university or community college before the advent of interactive voice response systems in the late 1980's, you will most likely will remember the days of course registration, recalled in the following paragraphs.

For first year students, course registration was a confusing process; not knowing which elective courses to take, especially if you had to decide in the following year, what you really needed in order to specialize in a discipline such as computers, psychology, or business. For those who survived their first year, it became a game to be able to find the best classes and professors, to meet your scheduling requirements. Not only was there a lot of stress involved in making these decisions, it was very expensive, and time consuming for professors, registrar's office staff, and for the students.

You could almost sense the impending stress of registration on campus, once the university had published its course calendar, and mailed out your registration forms. These forms sometimes had preprinted adhesive labels that indicated the courses in which you could potentially register. These labels would be placed on the registrar's course sheets, once you had successfully registered. Not all the labels could be used, only a subset would include the courses you wanted. For those courses which were prerequisites, you were not automatically assigned a course and class. You had a chance to select which class to attend. Some professors are better liked than others, or better course times might be more desirable. There would be a rush to fill those classes first.

Depending on the time of year, and the size of the university, the registration process could entail hours or even days. All you did was make your decision about which courses to take, and which professors you wished to have teach them. You would run around the different portions of the campus going from department to department, waiting in long and interminable queues to have your course assignments confirmed, or even rejected, because a prerequisite in the calendar was not published.

Even worse, one or more of the classes of your choice could be nearly full. To your great disappointment the faculty member, taking the registration for the department, would review your course history and course requests, and then ask you to go to the department chairperson for special permission to take that course. You would hunt down the tired and beleaguered chairperson, plead and negotiate your case with undue emphasis and theatrical pathos. If you were lucky, or the chairperson was very tired of hearing the same sad story 50 times before in the past 2 hours, you might then be given the required "blessing". You go back to the department's course registration queue, only to find out the class you wanted in the first place is fully booked, and now you have to take it next year. You would grumble, finalize your registration, and hand in your paperwork to the registrar's office.

There were the inevitable mistakes that sometimes crept into the system, and these had to be adjusted later by the students. This would further disrupt their well laid plans. Some students would even push the academic rules to the limit. They would purposefully book a course, only to drop it within the time limits the Dean of Academics would allow during the actual semester. They would switch it later, in order to get the course time or professor they originally wanted. They would gamble that some other student would do the same and drop out of the course that you were after.. Ahh... the fond memories of our days of our higher education! [27]

That was then. This is the here and now. Interactive voice response systems for course registration have evolved, from dedicated proprietary technology systems of the late 1980s, to open architecture, high volume integrated database system technology, that can handle thousands of calls per hour. York University has seen the development and utilized this technology over the years.

York University is one of Canada's largest universities, with a student population of over 40,000 enrolled in undergraduate and post graduate courses. Its sprawling campus is located in the Metropolitan City of Toronto, Canada.

Brent Wade, Senior Support Analyst, for York University MIS department, remembers the first student registration system implemented in 1989. "The demands of the system were such that interactive voice response technology was still evolving." The architecture was somewhat proprietary and eventually the University's needs surpassed its capability. The next generation of technology was purchased from IBM using Syntellect Technology, connecting to the mainframe. This has recently been succeeded by a DirectTalk/6000 system integrated with new client-server Student Information System. This had the growth potential and capacity to meet the university's requirements. The current usage of the technology has migrated to a client-server format, providing further expandability and flexibility at better costs.

System Overview
The technology comprises a IBM DirectTalk/6000 system with 48 telephone ports using T1 cards. It is a midrange IBM RS/6000 system. The DirectTalk/6000 manages the course registration

requests. It interfaces through the university's network, to a Digital Equipment Corporation VAX and UNIX systems which run an Oracle database that maintains the student registration system. This then allows up to 48 concurrent accesses to this database. In this configuration the DirectTalk/6000 system acts as a transaction processor against the Oracle based Student Information System (SIS).

As you can well imagine, the complexity of knowing the current status of a student in their degree, and knowing what courses are allowed to be taken, by this particular student in this point in their academic career, can be quite variable, and complex. There are two possible ways to handle these business rules. You could build all the rules within your interactive voice response logic, or you could store these rules within the database itself, and then have the database applications take action based upon those rules. The latter is much more preferable than the former, because this allows the database, along with the raw data about a student to coexist, and be easily changed by another Oracle application.

The business rules for the database are based upon year levels and the number of credits the student has successfully completed. When the interactive voice response application reads the results of applying the rules, it then knows what type of voice prompts to provide to the caller. This way there is consistency in the presentation and access to the course registration system.

If you coded the business rules inside your interactive voice response application, then a change in a business rule in the next semester, or the following year in a degree and its credits, could cascade into an immense coding and testing task. York University manages this complexity by building the business rules within the database application (SIS), and then having the interactive voice response application read those rules as it executes.

Call Flow

The students receive a notification by mail of the date and time of a registration window in March. The registration process starts in

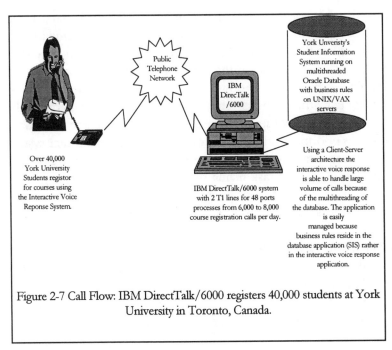

Figure 2-7 Call Flow: IBM DirectTalk/6000 registers 40,000 students at York University in Toronto, Canada.

Mid June and ends about 6 weeks later at the end of July. By that time about 80% of the Students have been registered. The system allows students to register for courses from any touch-tone phone. The window they have is 2 hours in length. This means they must be able to call the system within that time frame in order to reserve their courses. If they miss the registration window they have been assigned, they are allowed to call back the same day for a very small period of time, or call back on the following Saturday. By the end of August up to 90% (36,000) of the students have completed their registration using the system. From August to the first week of September, there is a designated free flow period where course add, changes, and deletions can be made on a first come first serve basis to the student body.

During the registration period, there is a constant and large influx of calls into the system every day. During the free flow period in August, the system has large spikes of calls.

Within a 14 hour day of operation the system will have taken between 6,000 to 8,000 calls. One reason for such a volume of calls is because of the "multithreaded" design of the application, and of the tasks to request data in the Oracle database. Multithreading is a technique that breaks down a single large request into its component smaller parts, and allows those smaller parts to be submitted to the computer system for execution. These tasks have different priorities and other factors that keep the computer system from being totally tied down by a single task. The computer quickly moves from task to task giving the impression of processing more information faster. Other design factors in the application have allowed this system to handle the volume of calls.

Business Analysis

The focus of the business analysis in this case is an obvious simplification of the registration process for the students as well as for York University. Besides these obvious benefits there is a return in better management of the system.

RETURN ON INVESTMENT

From the University's point of view the use of an interactive voice response system has reduced the costs and hassles of the registration process. The costs are reduced manpower and time to set up registration, as compared with a manual process. The process of embedding the business rules greatly simplifies the total flow. This then allows for applications that send out the notices by mail to the students, and as well effects the interactive voice response system and how it relates to the system.

Another and not so obvious saving is the management of the application. As Brent said, "anyone can write an interactive voice response system", the ability to manage it and be proactive about the management of the system is very important. One of the benefits of using the IBM DirectTalk/6000 platform is its ability to help proactively alarm potential problems within the system, before it has a major show stopping situation. This saves a lot of time and money for the university, especially during the system's heaviest usage.

From a business point of view of the evolutionary development of the interactive voice response system at York University, Brent has noticed an interesting management issue, in terms of skill sets that the staff need to have to successfully manage such systems. In the late 1980's the jobs of management of the systems were telecommunications specialists, hardware specialists who looked after the proprietary telephony hardware, computer hardware and software application specialists.

With each new generation of interactive voice response technology that has been adopted at York University, the skill sets have been concentrated in fewer and fewer staff members. The current generation at York University now adds the dimension of client-server databases using Oracle. The database application has migrated from a mainframe environment, to a mixture of smaller and more powerful machines. The staff that is required to manage this has been reduced in this configuration. Nonetheless, the remaining staff have to be well cross-trained in client-server computing and networking, remote database access, and in telephony. It is important to have the necessary training for the staff and the ability to hone their skills, so they can quickly react to newer situations. This does not minimize or negate the benefits gained from such an architecture and capability of the system.

One of the motivating factors to change to the newest form of the technology is that the legacy based systems were becoming prohibitively expensive to maintain. Migration to this new IBM architecture has reduced those expenses. Brent estimated that the reduction of costs were about 40% to 50%, from the previous generation of technology.

Besides the management issues, the registration process is available for long periods during the day and night. The business rules for the registration within the system provide a level playing field for all those who register. For those students who want changes, the opportunity to manage their schedule has an equal chance with others, during the free flow in the month of August. Faculty time is better spent looking after the exceptions to those rules, rather than being involved in the management of the registration process.

RIPPLE EFFECT

The benefits are to the customers, in this case the 40,000 students at York University. They do not necessarily have to waste hours or days getting approvals for courses. Students can register from almost anywhere and do not have to travel to the campus for the registration process. This allows students to register even from long distances away from campus such as out of province or even out of the country.

As to the number of hours saved per student, and the number of total man-days this represents could not be calculated at this time. It would be safe to say at least many thousands of hours of students time has been saved in the process of the student registration system, using interactive voice response.

Vital Statistics

The "Registrar" - York University	
Cost	*Statistics*
Weeks of labor to install system	Current system - 3 months
Initial System Cost	not reported
Est. Ongoing support Cost	reduced
System Size	
# Ports - Total/# Fax.	48 ports
# Users	400
# Calls/month	approx. 150,000
Computer	IBM RS/6000
Operating System	AIX
Industry	*Education*
Estimated Return on Investment	not reported
Cost Savings	
Manpower equivalency per year	
Estimated Manpower savings per year	
Reduction in costs from previous legacy system	40% to 50%

The "Registrar" - York University

Cost	Statistics
Earnings	
Generated new income/year	not applicable
Ripple Effect	
Estimated cost Saving to customer base	Tens of thousands of hours of time
Vendor	**IBM**
Product	DirectTalk/6000
End User	York University, Toronto
Computer Telephony Technologies Deployed	
Interactive Voice Response	Yes

Table 2-5 The "Registrar" - York University - Vital Statistics

---

[27] This description of manual course registration is from the author's own recollection of how it was done while attending Wilfrid Laurier University, Waterloo, Ontario between 1978-1981, & 1983-1984. This is not how it was exactly done at York University. The system would be much like it. The purpose of this section is to show the difficulty of the manual system, rather than the accuracy of how it was actually performed at York University.

'Investments' by the Nile - National Bank of Egypt

W hat are the challenges of deploying computer telephony technology outside North America and Europe, where we have very modern and easy access to telecommunication services? What are the cultural issues, and liaison that is necessary for success? MediaSoft Telecom's work with the National Bank of Egypt has been a success by using their IVS product to provide voice and fax services for this customer.

Figure 2-8 MediaSoft's IVS Builder & Server helps National Bank of Egypt 's customers with 'investment' management.

Background

The National Bank of Egypt is the largest bank not only for that country, but as well in the Middle East. Although government owned, it is governed by market forces, such as competition. The bank has over 300 branches, of which there is one in New York City, and London, United Kingdom. They have a large number of customers outside the country, such as in Saudi Arabia and others in the Gulf region.

The bank has a tradition of keeping a competitive edge in the market by providing to its customers the newest technologies to help service their needs, and to bring new products to market. For example, the National Bank of Egypt was the first bank to provide automated teller machines, and to offer VISA credit services.

In accordance with this tradition, the National Bank of Egypt had also provided videotex terminals and PC based videotex access to their customers. This is where MediaSoft had been originally involved some years ago. During that time they had developed the access methodology to the database server. This was later expanded to support the voice and fax services, using MediaSoft's IVS system.

CULTURAL ISSUES

There are cultural issues that needed to be addressed:

1. They had to accommodate Arabic as a language for the system prompts. Unlike Saudi Arabia which does not permit females in business, and hence female voice applications, the National Bank of Egypt could record both female or male voice prompts. Additionally, the Arabic language demands that dates and amounts are announced differently than in the Latin based languages such as English. There are special challenges when faxing information, since the Arabic language is read right to left, rather than left to right, so the formatting of text is a special challenge. Moreover, Arabic allows for a single letter in the alphabet to have up to four different shapes. The shape is dependent upon its usage within the context of the word. This requires special context sensitive analysis of the construction of a document that is faxed.

2. An important point to consider is that Islam does not permit one to earn "interest". This is also incorporated into Egyptian secular law. The bank can only allow its customers to make "return on investments", rather than "interest". This affects the application menu prompts both in written form as well as in voice prompts for the interactive voice response applications. So you would see things such as "rate of investment" rather than "interest rates." A bank customer's deposits are represented in investment certificates.

3. Support staff have challenges in supporting this account because of the 7 hour time difference. This means that they have to get up early to work with them. Moreover, because of the cultural differences and environmental differences of Cairo, the support staff that make the trip there are not as eager to do so on the next opportunity.

TECHNICAL CHALLENGES

The telecommunications infrastructure is not as strong as in western countries. There is a preponderance of rotary dial phones, with only 5% of the country supporting touch-tone. This is a special challenge, since interactive voice response applications require some mechanism to determine what number has been selected from the telephone keypad or dial. This was solved by providing pulse-to-tone detection equipment from Telelaison. This technology converts the pulses that a rotary dial phone makes into touch tones, so that the IVS system could understand what the caller had dialed.

System Overview

National Bank of Egypt has deployed 3 applications:

1. MediaSoft developed for National Bank of Egypt a commercial accounts banking by phone system, that provided:

    a) Obtaining the latest "investment" rates

    b) The latest exchange rates. This is of importance, since it been only in recent years that the Egyptian Pound was allowed to be convertible to hard currencies, such as US Dollars.

    c) Transfer money between account(s) for the same customer.

    d) Transaction history on the account(s).

    e) Leave messages for the bank for such things as ordering more cheques, or other notices.

    f) Fax -on-demand service to send account details by fax to a phone number on file.

    Of interest, was that this IVS application was based upon the same access methods as the videotex, and PC access applications to the mainframe database. This simplified the access and the security for the bank.

2.   Another application was the result of the desire of the government to encourage its citizens to invest in the banks. The reason for this is based upon the culture of the country, and especially to those that live in the regional areas outside Cairo. Culturally, and traditionally people would always have their money at home, or with themselves. They would feel much better if they were able to see it and be with it, rather than have it invested somewhere else.

In order to encourage investment, the government created a lottery for those who deposited their money at the National Bank of Egypt. Because of the aforementioned religious and legal issues, a depositor would receive an investment certificate. Each investment certificate would have a unique number. Once a month the government would randomly select a certificate number. The owner of the certificate would receive another certificate of the same value for free.

This generated thousands of inquiries every day. The number of calls posed a problem in that it tied up value resources. Because of the call volume, the bank had to have 8 to 12 employees just to answer these calls. As well, this staff could only answer the phone during National Bank of Egypt working hours. This required a lot of line utilization during that period. The IVS application did the following:

a)   A caller would call and enter the certificate number, and the system would inform them that they had won, or that they should try again.

b)   The benefits were that the IVS system was able to allow National Bank of Egypt managers to deploy the 12 people for more productive work in the bank.

Because the IVS system was available 24 hours a day, the number of lines needed was reduced. This may seem an insignificant benefit, but in other countries outside he industrialized world, telephone line access is a precious commodity, and it may take many months or even years to obtain phone lines.

3.   The other application that National Bank of Egypt developed themselves, using IVS, was for their VISA services. The system allows customers to determine their balances and amount remaining of their credit line, and as well to provide information on payment history.

Call Flow

The call flow is relatively simple to comprehend. However the telecommunication environment dictates that analog telephone lines with a preponderance of rotary dial versus touch-tone

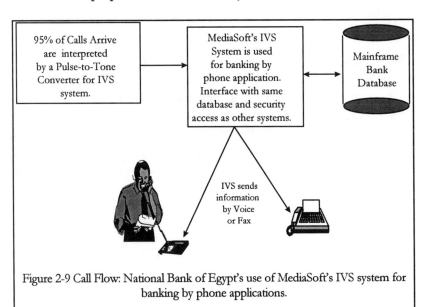

95% of Calls Arrive are interpreted by a Pulse-to-Tone Converter for IVS system.

MediaSoft's IVS System is used for banking by phone application. Interface with same database and security access as other systems.

Mainframe Bank Database

IVS sends information by Voice or Fax

Figure 2-9 Call Flow: National Bank of Egypt's use of MediaSoft's IVS system for banking by phone applications.

required pulse-to-tone conversion. Moreover, the system had to be efficient in terms of line utilization.

Business Analysis

The focus of the business analysis is the competitive edge and corporate image that was of prime importance to the National Bank of Egypt, as well as the ability to provide better service for customers at lower cost to the bank.

RETURN ON INVESTMENT

The major return on investment for the National Bank of Egypt is that it has been able to obtain a competitive advantage in its market place. At the time of the interview in October, 1996, the National Bank of Egypt was the leading bank providing bank-by-phone/fax services in Egypt and by nature of its size, probably in the Middle East. As Bachir Halimi, president of MediaSoft Telecom said, "it protected the bank's leadership in the market, preserved and increased its corporate image to its customers, and provided better customer retention, and increase in the number or customers."

RIPPLE EFFECT

The ripple effect can be measured in the quality and ease of international service for the National Bank of Egypt commercial customers. Moreover, it shows the power of interactive voice response systems to do not only reduce the number of staff to answer calls (especially the certificate lottery staff), but also to be used as a tool to encourage customers to use banking services, and help them to gain access to international banking standards and services.

Vital Statistics

'Investments' by the Nile - National Bank of Egypt	
Cost	*Statistics*
Weeks of labor to install system	Commercial Accounts IVR - 1.5 months at MediaSoft and 1.5 months in Cairo, Egypt.
Initial System Cost	
Est. Ongoing support Cost	
System Size	
# Ports - Total/# Fax.	Initial system consisted of 36 ports (comprised of 3 Dialogic D120/D cards with one port fax card.)
# Users	tens of thousands
# Calls/month	135,000
Computer	Pentium
Operating System	SCO OpenServer
Industry	*Banking*
Estimated Return on Investment	not reported
Cost Savings	
Manpower equivalency per year	12 man years minimum
Estimated Manpower savings per year	
Reduction in telephone costs/year	better line utilization.
Earnings	
Generated new income/year	not reported
Ripple Effect	
Estimated cost Saving to customer base	
Vendor	*MediaSoft Telecom*
Product	IVS Builder & Server
End User	National Bank of Egypt
Computer Telephony Technologies Deployed	
Interactive Voice Response	Yes
Fax On Demand	Yes
Voice Messaging	Yes

Table 2-6 'Investments' by the Nile - National Bank of Egypt - Vital Statistics

Chapter 3

Fax Servers

Fax server technology on UNIX provides a network wide, multiplatform access solution for automatic faxing of office documents, broadcast faxing for marketing, distributing drug test results, or helping with weather reporting and forecasting. Fax server technology, can also be further leveraged, by combining it with an interactive voice response application generator, to create a special genre of interactive voice response called Fax-On-Demand, as we have seen in the chapter on Interactive Voice Response.

UNIX shines as a platform for these applications, since the need for high reliability and scalability of the platform, and of the application, can make these applications very valuable and mission critical. True to the traditional role of UNIX, playing the "glue" of disparate networks, platforms and applications, UNIX based fax servers allow many different kinds of users, and computer operating systems to launch faxes. A UNIX user using X-Windows, world wide web browser user, MS Windows95 user, and a character based terminal user, are able to launch faxes from their computer or terminal. The application software in each environment merely has to redirect the printed output to a UNIX "fax printer", rather than to a regular UNIX print queue. So if you are a MS-Windows user of Microsoft Word and you need to print to the fax printer, the document is converted to a graphic file type (usually Adobe PostScript™) and sent to the UNIX server for conversion to a fax format. It then is faxed from the UNIX system, via a regular modem with fax capability, or sent out via a high density fax card inside the fax server The computer generated fax is much clearer to read since it is not a scanned image of the original. This allows different users to send different types of documents from different environments with little or almost no user training.

In our review of the case histories below, they are based on typically lower density fax modem applications that have integrated fax server software technology to increase the productivity. They show the tremendous cost savings, and a relatively inexpensive and easily installable technology. This begs a question. Where are the case histories about high density fax board server technology?

The reason for the exclusion of the high density fax server technology in the case histories presented in this chapter is more an accident of fate - sample bias - rather than of purpose. The bias in the case history compositions are based upon the responders to our request for case material, and our original sample to solicit. We happened to be more fortunate to gather information in one category, than in another. Please read the following case histories. Then refer to Technical Reasons for Using High Density Fax Servers on page 151 of this chapter, to understand where and why these products are preferred in large fax server environments.

The "Hurricane Tracer" - Universal Weather & Aviation

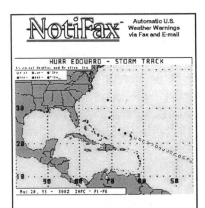

Figure 3-1 V-Systems helps Universal Weather & Aviation to serve their customers with customized bulletins via fax without customer intervention.

I f you were an insurance or utility company, would you not want to know where impending weather disasters and losses are about to strike? Hurricane watches & warnings for tornadoes, thunderstorms, ice storms and flash floods are some of the perils that would be of interest to insurance and utility companies. Planning ahead for such events allows them to be more responsive to the needs of their customers. For aviation and navigation planning special weather statements such as heavy rain, winds, and fog conditions could be very valuable information. How can this information be quickly transmitted, and without any specific request of the customer requiring it? The answer is a fax server. In this case history we examine how Universal Weather uses V-Systems VSI-FAX software to transmit thousands of faxes per day to their customers.

Background

The history behind selling aviation weather by Universal Weather & Aviation had its beginnings in the United States following the Korean War. The United States military decided to sell their smaller and no longer needed transport jets to corporations. These sales were not to commercial airline carriers. As part of the deal, the US military offered to supply the peripheral crew services to the corporation's own crews. This being the same services a airline would be provide to their crews. One of these services was weather information. Unfortunately, over time, the US military found that providing these services became too expensive to deliver to the corporations who had purchased the aircraft. They then moved out of that business.

This left an opportunity, for Tom Evans, the founder of Universal Weather and Aviation, in 1959 to start a private weather company. The mission of the company then was to gather the information pilots needed and resell it to them. This service has been the mainstay of the company for many years. Universal Weather & Aviation has grown to be a much more comprehensive service providing company. It has 30 offices around the world, supplying information to several thousand customers, who in turn fly several thousand aircraft. The company is a world leader in the aviation industry sector, called corporation aviation service management. The services provided go far beyond providing weather information. Universal Weather & Aviation provides legal documents such as visas, passports, and fly-over permits for crews. As well, they provide ground transportation, discount accommodations, contract for fuel, providing credit for aviation fuel at several thousand airports around the world. Beyond this, the most important services is still providing weather information to the corporate air crews.

The information that is being sent to Universal Weather and Aviation's customers is not of the same nature that one may generally acquire from the US National Weather Service. Any person can call the US National Weather Service and demand weather information. The information they receive is provided in a structured format that may not be as useful or as timely to certain types of customers. For example, it is important for a pilot to know the wind speeds, and wind sheer problems around a thunderstorm to avoid trouble. He would want to know about rain conditions on the runway at the destination airport. Notification of localized and dangerous weather by state, county, and city, and in what format of presentation is the added value the company provides. These are examples of how specialized are the kinds of weather bulletins provided by Universal Weather & Aviation. This information may not be readily available from the US National Weather Service.

In fact, the US National Weather Service, and other 3rd party weather companies provide raw data and information to Universal Weather for interpretation by their 60 meteorologists, and how they distribute that information is the main business of the company. The evolution of communicating this information has been by telephone, then by telex based services, proprietary computer networks, and now by fax, and Internet email.

One of the advantages of the newer technologies such as Doppler radar information from the US National Weather Service, has allowed Universal Weather to broaden their customer base from the aviation industry to other industries such as insurance companies and utilities. This has allowed for the creation of the NotiFax product.

NotiFax is a weather information service product which allows a customer, for a small monthly fee (starting from $40/month) to receive by fax, weather bulletins up to 7 days a week, 24 hours a day, without any intervention or action by the customer. Although the format of the individual bulletins are not customized, the number, type, and nature of a bulletin can be customized for the customer in a customer weather profile, which is stored in a database. Fred Rogers, of Universal Weather & Aviation, said, "Snow, wind, heavy rainfall watches and warnings, tornadoes, and a whole spate of hurricane watches and warnings, which appeal mostly to the coastal companies", are examples of the kinds of information that can be associated with a customer's profile. .After the data from the different weather sources has been analyzed by the meteorologists, or a weather advisory for a specific area is generated by the US National Weather Service, the information is then cross tabulated with the requirements of the customers profiles, and the appropriate information is prepared and scheduled for faxing or by email.

NotiFax is comprised of 2 kinds of information. Most of it is textual, in some other cases such as hurricanes, it also has an

Special Weather Statement

WWUS35 KNYC 142041

SPSNYC

NYZ067>081-NJZ002-006-011-CTZ005>012-150300-

SPECIAL WEATHER STATEMENT

NATIONAL WEATHER SERVICE NEW YORK NY

445 PM EDT MON AUG 14 1995

...A HEAVY SURF ADVISORY IS IN EFFECT FROM MANASQUAN INLET/NJ TO MONTAUK POINT/NY.STRONG RIP CURRENTS ARE LIKELY AND MINOR TO MODERATE COASTAL

FLOODING IS POSSIBLE ALONG LOCAL COASTAL BEACHES...

EVEN THOUGH THE CENTER OF HURRICANE FELIX IS WELL OFFSHORE /NEAR BERMUDA/ ...THE HURRICANE HAS GENERATED LARGE SWELLS WHICH CAUSE DANGEROUS SURF CONDITIONS...INCLUDING STRONG RIP CURRENTS.

EARLIER TODAY...THE OCEAN BEACH POLICE DEPARTMENT IN SUFFOLK COUNTY/NY REPORTED SEA SWELLS UP TO 8 FEET WHICH CAUSED BEACH CLOSURES ON FIRE ISLAND. SIMILAR REPORTS FROM THE U.S. COAST GUARD AT JONES BEACH IN NASSAU COUNTY/NY WERE RECEIVED.

UNFORTUNATELY...BECAUSE HURRICANE FELIX WAS EXTREMELY LARGE IN AREA AND WAS MOVING SLOWLY NORTHWEST TOWARD THE MID-ATLANTIC COAST...THESE CONDITIONS WILL CONTINUE TO DETERIORATE DURING THE NEXT FEW DAYS. GRADUALLY BUILDING SEA SWELLS COMBINED WITH HIGH TIDES WILL ALSO CAUSE MINOR TO MODERATE BEACH EROSION AND COASTAL FLOODING. IF THIS TREND CONTINUES...A COASTAL FLOOD WATCH FOR LIFE THREATENING COASTAL FLOODING MIGHT BE ISSUED SOON.

THE PATH AND INTENSITY OF HURRICANE FELIX IS STILL UNCERTAIN. HOWEVER...THIS IS A LARGE AND FORMIDABLE HURRICANE. FURTHER STATEMENTS WILL BE ISSUED AS NEEDED.

GC

Figure 3-2 Sample NotiFax: Special Weather Statement. Courtesy of Universal Weather & Aviation. Used with special permission

accompanied map showing the previous, actual and projected track of the storm over the globe. The key feature of this service is that it does not require any action on the part of the customer to obtain the information. The information is very quickly sent out after its creation. Customers can request the information during business hours only, or by 24 hours a day 7 days a week. Of interest, the majority of NotiFax clients are not aviation related. Transportation companies, utilities, and insurance firms are examples of industries that use NotiFax. Fred Rogers, said "you wouldn't believe who really wants weather information."

System Overview

Information from various sources are collected and analyzed by the meteorologists. Meteorology is a data and compute intensive task. Large databases of information and heavy use of graphics requires a lot of computational power, which have traditionally been provided by UNIX based workstation manufacturers. At Universal Weather & Aviation, the weather analysis and information dissemination are performed on Hewlett Packard (HP), Sun Microsystems, and IBM RS/6000 workstations. V-System's VSI-FAX product is loaded on all three kinds of platforms. The systems work together in a network. After the weather has been analyzed, the meteorologists will view the potential output of the NotiFax hurricane maps on their HP workstations, using the X-Window based VSI-FAX fax viewer that is supplied by V-Systems as part of the product. X-Windows is an open standardized windowing technology that allows applications to be executed on different platforms, without having to write specific software for each platform. In this way, one can see the potential output of a fax from any workstation that has the X-Window software installed on the system. VSI-FAX supports viewing and creation of fax documents, from character as well as MS-Windows interface.

Once a bulletin and optional maps have been identified to be sent to the customer's profiles, the data is then moved to the Sun Microsystems SPARC 10 for further processing. The usual output from the weather applications on the HP workstations is in a postscript format. Postscript is a graphics page description language developed by Adobe, and is used to display text and images in a graphical layout. Many software packages on the market today support Postscript, as do many laser printers. A postscript file is a large character based file that is full of numerical information and

Postscript commands. A conversion process is done on the Sun Sparc 10 workstation from Postscript to fax file. The Sun Sparc 10 is a machine well known for its ability to provide a lot of computing power, which is needed by Universal Weather & Avaiation, since they are converting many Postscript files to fax files every hour.

Converted information is then sent, along with the customer's fax destination number, to the IBM RS/6000 machines which have up to 60 Multitech fax modems attached. These machines are dedicated to fax the very large amount of information that is required. NofiFax typically uses 25 of the 60 fax modems, with a backup of another 20 modems, in case of failure of one of the IBM RS/6000 machines. The remaining ports are being used for other services and regular day to day administration communications. Depending on the time of year, such as hurricane season, and how active the weather is, the NotiFax system can generate as little as 2,000 faxes or as many as 9,000 faxes per day. Using VSI-FAX's scheduling, prioritization and managed queues, Universal Weather will be able to easily manage the modems services. Because of the variability of the demand on the system, the need to have a large bank of modems available makes the system relatively inexpensive to manage. There is no additional software costs, only relatively small modem costs.

Steve Perkins of Universal Weather & Aviation said, "One of the biggest benefits our customers say about our NotiFax product is its clarity of the images and text. We do not loose any clarity because the documents are not scanned in by a regular fax machine." The is achieved because the images and text are completely computer generated, and faxed directly as computer data.

Call Flow

The call flow for the use of the fax server is easy to understand. Universal Weather & Aviation has developed their own software to generate their added value weather reports and delivery of 3rd party weather bulletins. Once the data is ready for fax transmission, it is prepared for transmission on a dedicated Sun Microsystems Sparc 10 server, and sent to the IBM RS/6000s to be faxed using VSI-

FAX. The customer's fax machine is automatically dialed, and if necessary retried if busy, or no answer, and the NotiFax document is then delivered.

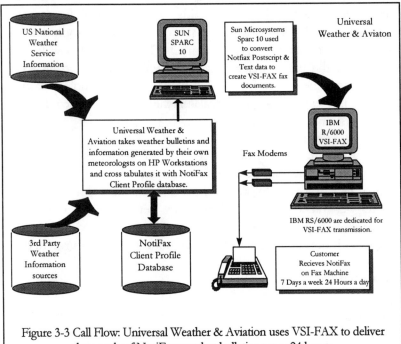

Figure 3-3 Call Flow: Universal Weather & Aviation uses VSI-FAX to deliver thousands of NotiFax weather bulletins every 24 hours.

Business Analysis

In a very competitive marketplace, where the product, which is weather, can be made available from various other sources, added value such as customization of the type of bulletins, and the efficiency of its delivery and customer service are a sustainable competitive edge. NotiFax provides a new market for Universal Weather & Aviation, and is reaping the rewards for its services. Because of the competitiveness of this market, some of the questions on return on investment and ripple effect can not be discussed at this time. However, general statements can be made, which will not jeopardize their position in the market.

RETURN ON INVESTMENT

The return on investment for Universal Weather & Aviation is measured in new market and edge. By using a UNIX based fax server technology, they were able to leverage their existing knowledge of UNIX and data resources, and apply sophisticated added value to service this new market. This meant a lower cost of entry into the marketplace. Given the volume of fax data being transmitted, the ability for the UNIX systems to handle the large volume is a testimonial to the scalability and as well as the reliability of the system.

Since they are transmitting thousands of documents a day, an interesting statistic can be ascertained. A known assumption that it would take at least 5 minutes for a person to fax 1 document, and there were at least 2,000 documents per day, and a fax was successful on every first attempt, then the VSI-FAX system is providing equivalent of about 28 man hours each 24 hour period.

RIPPLE EFFECT

The benefits to the NotiFax customer is that they do not have to assign staff to research the information they need. It is automatically sent without any intervention. This saves them valuable time in order to service their customers.

Vital Statistics

The "Hurricane Tracer" - Universal Weather & Aviation	
Cost	*Statistics*
Weeks of labor to install system	not reported
Initial System Cost	VSI-FAX software for each IBM RS/6000 $2,000
Est. Ongoing support Cost	not reported
System Size	
# Ports - Total/# Fax.	25 fax ports
# Users	up to 300 internal VSI-FAX users
# Calls/month	at least 60,000/month

The "Hurricane Tracer" - Universal Weather & Aviation

Cost	Statistics
Computer	HP, Sun, IBM workstations
Operating System	UNIX operating systems: HP/UX, Solaris 2.4, AIX
Industry	*Aviation & Meteorology*
Estimated Return on Investment	not reported
Cost Savings	
Manpower equivalency per year	VSI-FAX provides the equivalent of 1275 mandays per year ((28 hours per day x 365 days per year /8 per man day))
Estimated Manpower savings per year	not estimated
Reduction in telephone costs/year	not estimated
Earnings	
Generated new income/year	not reported
Ripple Effect	
Estimated cost Saving to customer base	
Vendor	*V-Systems Inc.*
Product	VSI-FAX
End User	Universal Weather & Aviation
Computer Telephony Technologies Deployed	
Fax Server	Yes

Table 3-1 The "Hurricane Tracer" - Universal Weather & Aviation - Vital Statistics

The "Pleasure Craft" - Stingray Boat Company

Figure 3-4 Stingray Boat
Increases customer
communications, and reduces
labor costs using Faximum.
Stingray Boat Image: Copyright
1996 Stingray Boat. Used with
special permission.

Manufacturing pleasure craft is a complex business, requiring a lot of communication to suppliers and to customers. Stingray Boat, of Hartsville, South Carolina, shows how Faximum software has saved the company time, money, and increased customer communications using an integrated UNIX based solution.

Background

Stingray Boat manufactures motorized pleasure boats that start from $10,000 (retail price) and range an outboard motor boat 4.57m (15ft) long to 7.01m (23ft) with all the accessories. Stingray Boat has 100 employees, with a worldwide dealer network of over 80, with no direct sales to the public. Support and communication to the dealers is a key factor in the success of Stingray Boat. Stingray Boat builds their boats on a per order basis, so as reduce costs in inventory and other materials. This makes the efforts to support the dealer with current and accurate information even more necessary, since the product is being manufactured specifically for the dealer.

Before the fax server was installed and integrated into the application software, there was an ever increasing delay and line up for use of the fax machine at their office. When the purchasing agent for the company had to send 3 purchase orders, they had to be created in the Filepro system, printed, and then a cover page added. When the purchasing agent wanted to send the fax, he often had to wait in queue to use the machine. Then he would have to spend another 5 to 10 minutes faxing the documents he had generated. It was not unusual for many of the staff to wait 10 to 15 minutes to get access to the fax machine. Stingray Boat has over 200 vendors, of which any number of those would need to fax purchase orders for a single boat order. Moreover, the cost of transmission of these documents happened during regular business hours. This meant that all transmissions were being made during peak business hour rates. This was unnecessary for some of the transmissions, which could be sent at lower rates at night.

When fax machines became prevalent in the late 1980s, sales leads to the dealers would be printed off, and then manually faxed. This involved a single employee spending a significant portion of her day sending faxes to the dealers.

System Overview

Robbie Gilbert, Information Systems Director, described his system as being a complete enterprise solution using an in-house built Filepro database system. This resides on a SCO UNIX 3.2.4.2 system with 44 terminals and 11 printers, running on a Pentium based personal computer. Faximum software with a single Multitech modem for faxing provides the needed capability. The Filepro application software provides complete order processing, vendor and purchase order management, inventory management, manufacturing process control and shipping information.

Robbie identified four main areas where automated faxing has made a significant contribution in terms of time and money.

Generated leads come from different sources. These leads are then entered in the Filepro system. A custom letter is generated and sent to the prospect. Then the nearest dealers to that one customer are also notified by fax about the prospect in their territory. The decision on which dealers are notified is based on several criteria, such as location of the prospect to the dealers. The prospect also gets the same list of dealers.

Once the dealers complete an order form, they fax it directly from the paper fax machine in their office. The order entry staff then enters the order into the system. Each boat may have some special features or other parts that need to be identified with the boat. Once the order has been placed in the system, a confirmation of receipt of the order is then faxed back to the dealer, verifying the make and parts of the order. Each order confirmation consists of 2 to 4 pages, which includes all the boat's options, shipping date and expected delivery date. Since the order confirmation is faxed directly from the computer system, the quality of the printout at the dealer's fax machine is near laser printer quality. This clarity is achieved because the document is not manually scanned by a fax machine.

After an order for a boat has been placed in the system, the purchasing agent has a bill of materials, which will need to be reconciled with inventory. The system does this reconciliation and consolidates the parts by vendor, and prints a materials requirements list of parts that need to be purchased to complete the boat. The purchasing agent needs only to review this list. He then uses the system to decide which vendors will receive purchase orders, after which they are faxed directly to the vendors. This provides immediate savings for vendor management processes.

A production run may have up to 6 to 8 orders being processed at a time. The order manager and schedule production person decide which boats are going into production. They fax a production confirmation of the order to the dealer. This provides the dealer with one last chance to make any changes to the boat's configuration before the product is built. The dealer then signs off on this confirmation, and faxes it back to the production department. On occasion a dealer will call back with a hot order change. This occurs when a dealer is given an opportunity to sell a boat which has a small feature change, after he has signed off on the production confirmation. The production manager then makes the change to the boat configuration in the system, and the system immediately faxes the change back to the dealer to help close the sale. The day the boat is put into production, all the parts for the boat are known, and some of these parts will have serial numbers on them. The dealer receives another fax providing details of the components along with their serial boat numbers. This is important to the dealers, since they may require this information to process any financing arrangements they have had to put in place for their customer. Finance companies usually require this information to be in the loan documentation. After the boat has been produced, the Stingray Boat sales representatives are notified by fax that a shipment of boats is being delivered to the dealer in that person's territory.

Another order process deals with extra parts from the warehouse. If a dealer wants to purchase a canvas cover for a boat or a quantity of spare parts, the order is processed and an invoice is faxed to the dealer. This shows the shipping and delivery of the products, and the number of parts that could be shipped. If the dealer wanted 10 pieces of a part, and only 5 were being shipped, the faxed invoice would reflect the 5 pieces on back order. Robbie Gilbert commented that this has eliminated a lot of calls from the dealers inquiring about the status of the parts orders. Each call would take at least 5 minutes to complete, and there were many of these calls per day. This has allowed the personnel in the order department to work on more important tasks, and solve more difficult customer problems. If there are any changes, or if the dealer does call to find out the status, the call is usually much shorter because the order personnel only needs to find the order in the system, and fax the status of the order with only a simple press of a button.

Since Robbie Gilbert has customized the integration of the enterprise database system with Faximum via UNIX shellscripts, he can carefully control the priority of the fax transmissions. For example, purchase orders and dealer orders are assigned a very high priority over all other fax traffic. Customer parts orders and invoicing for parts orders are also fairly high priority, and are usually delayed for faxing until the end of the business day. Another interesting feature of Faximum software's scheduling algorithm is to reconcile multiple individual fax documents to the same destination fax number within the fax queue. As different departments send information by fax to the dealers, Faximum examines the requests, and if there are two or more faxes to the same phone number it keeps the phone line open and then sends the subsequent fax documents on the same call. This feature does save Stingray boat money. It takes on average about 6 seconds to setup and then tear down each call. This is time that Stingray Boat does not pay to their long distance carrier. Robbie indicated this happens quite frequently every day.

Although most of the fax transactions are generated via the Filepro based manufacturing system, there is still a need for a limited number of individuals in management and sales that need direct access to the Faximum direct database. Each night the data in the manufacturing system updates the contents of dealers and other groups within the Faximum databases, which allow for group faxing to specific groups of individuals. If management wanted to send a message directly to all the dealers or sales representatives in Canada and the United States, management would type the message directly within Faximum, and then send it to a designated list of dealers. A mass fax with an attached file created with a desktop publishing product can also be sent by this method.

Call Flow

The call flow is only outbound faxing spawned either by a dealer order, or by a management request for a broadcast.

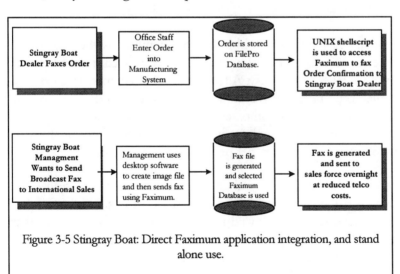

Figure 3-5 Stingray Boat: Direct Faximum application integration, and stand alone use.

Business Analysis

The following business analysis uses a published study for the quantifiable numbers for the basis of return in investment. This study showed the benefits of using fax based systems as compared to regular US Postage, and labor rates.

RETURN ON INVESTMENT

Although Robbie Gilbert had not completed a time study of the savings of the faxing solution for his company, there is a generally accepted cost saving structure for fax servers in the computer industry. Depending on the time of year, Stingray Boat generates anywhere from 2,500 to 3,500 faxes per month, with an average of 2 pages per fax. This on average for the year is about 3,000 faxes per month. This translates to about 115 in low season and 160 faxes per day in high season. Each document costs about $1.79 mailing and labor as a mailed document. Faxing the same document costs $0.50 per fax for a savings of $1.29 in 1994 dollars[28]. Therefore, the estimated savings for Stingray Boat at the average monthly volume is $3,870 or $46,440 per year.

Another major factor Robbie Gilbert identified is the amount of time saved because of automated fax retry by Faximum, whenever the destination fax is busy. Although modern fax machines do have a memory storage for scanning a document, and then provide retries, this does not compare to the fax server's ability to handle a large number of simultaneously requests and the unattended management of the number of retries the fax is sent.

Most of this saving is in labor time. It is important to realize the savings has not resulted in the elimination of positions at Stingray Boat. Instead, it has allowed those key staff members to do other tasks which help support the overall efforts of the company to run more smoothly, and increase customer support.

RIPPLE EFFECT

One of the benefits for the dealers around the world is that they do not have to spend extra time talking directly to individual departments as their orders are processed, since faxing has reduced this time. Robbie Gilbert said this was most dramatically shown in the parts ordering function. He estimates that it has saved 5 minutes or more on each call as the dealer inquires about his order. Based on historical trends, the fax notification process saves on average of 31.25 man weeks for all the dealers' time each year. This is the lapse time, and does not calculate the estimated savings for telephone toll charges for these calls.

Vital Statistics

The "Pleasure Craft" - Stingray Boat Company

Cost	Statistics
Weeks of labor to install system	2 man months over a period of 6 years of progressive evolution and integration of fax solutions into the manufacturing system.
Initial System Cost	$1,000
Est. Ongoing support Cost	not reported
System Size	
# Ports - Total/# Fax.	1 Fax port (Multitech Modem)
# Users	8
# Calls/month	2500 to 3500 depending on season or on average 3000 faxes per month
Computer	Pentium PC
Operating System	SCO UNIX 3.2.4.2
Industry	*Manufacturing*
Estimated Return on Investment	$46,440 per year in labor efficiency, and reduction in postal costs.
Cost Savings	
Manpower equivalency per year	375 man days per year
Estimated Manpower savings per year	$45,000
Reduction in telephone costs/year	not reported
Earnings	
Generated new income/year	Not Applicable
Ripple Effect	
Estimated cost Saving to customer base	31.25 man weeks/year
Vendor	*Faximum Software*

The "Pleasure Craft" - Stingray Boat Company

Cost	Statistics
Product	Faximum
End User	Stingray Boat Company
Computer Telephony Technologies Deployed	
Fax Server integrated to application	Yes

Table 3-2 The "Pleasure Craft" - Stingray Boat Company - Vital Statistics

---

[28] Estimate savings are taken from pages 2-6 to 2-8, <u>Fax-on-Demand, Marketing Tool for the '90s</u>, 3rd Edition Copyright 1994 AB Consultants, a partnership 3250 McKinley Drive, Santa Clara, CA 95051; 1-408-243-2234

The "Robofax Broker" - Headlands Mortgage

I n the mortgage business, quick and efficient transmission of approvals or denials of loan applications to mortgage brokers is an important valuable service. The ability to provide other electronic means of communicating, such transactions provide a value added value and customer service, which brings these important customers back with more business. How can you better utilize the time of your staff to communicate this information? The answer is fax server technology. The way this has been implemented at Headlands Mortgage Company is very interesting technically, and saves the company a bundle in staff time and money.

Background

Although Headlands Mortgage does most of its business with mortgage brokers, they do sell mortgages directly to the public. "Headlands Mortgage Company (Headlands) was founded in 1986 by Peter T. Paul, President, and is head quartered on the San Francisco Bay in Larkspur, California. Since its inception, Headlands has originated over 92,000 loans, totaling over $14 billion dollars. We currently service over 28,000 loans for 4.3 billion dollars. Headlands primary focus is wholesale lending, with a client base of some 4,000 Brokers throughout the western United States. Headlands is noted for developing loan products which meet current home buyers' financing needs."[29] The company has 4 other offices in Southern California, 2 in Oregon, 1 in Bellevue, Washington, and another in Phoenix, Arizona. These remote offices are connected in a wide area network to a Data General Aviion UNIX system.

Each day they will have about 50 staff members throughout the company producing about 300 mortgage approval or denial letters. These documents consist of 2 parts. The first is the standardized boilerplate text, while the second is the data from their database. When a Headlands staff member makes a request of the MortgageFlex software to create a letter, the database information is merged with it. They always produce a paper copy of this document, and optionally, they are asked if they want to fax a copy of the document at the same time, if there is a fax number in the database for the mortgage broker. The document is created, one image goes to the printer, while the other image is sent as an email attachment for later faxing by the fax robot. If they had to fax it manually, this would take time away from doing other tasks, such as working on the next mortgage approval.

System Overview

Mortgageflex is a product developed by Morgageflex of Jacksonville, Florida, which Headlands has internally customized. It is located on a centralized host Data General Aviion UNIX computer system, in the head office of Headlands Mortgage in Larkspur, California. With a geographically dispersed company, email is a very important communication medium. This was integrated within the Mortgageflex system to provide easy communication between users. It was a logical extension to allow Mortgageflex to send email to another new specialized user, which in reality was not a person, but a piece of software residing on the same UNIX host computer.

This is a fax robot" which receives email messages, and then stripped out proprietary header information, and the body of the message. The body of the message could be a letter to a mortgage broker indicating an approval or denial for one of his customers. The relevant data from the Mortgageflex database is merged within the body of the text letter, and then sent as an email message to the fax robot user. Based upon the instructions within the header information, the message is converted to a fax and scheduled for transmission by "V-System's VSI-FAX.

This is not as a circuitous as it may seem. One may think it would be easier to just send a fax directly from the MortgageFlex software, rather than have it sent as an email message to be converted as a fax message.

There were two reasons for this. Firstly, they did not want to have to train users on yet another software package. Although V-Systems VSI-FAX comes with a comprehensive set of interfaces for both character based applications, X-Windows, and MS-Windows platforms, the need for them to learn these was not an

146

imperative. Secondly, the design allowed for some application independence of the fax platform. Since UNIX already comes with a sophisticated mail system with every copy of the operating system, then it made sense from a system architecture point of view, to use this as the means of messaging between users and between applications. The proprietary header scheme allowed for other potential programs to submit faxes to the system, and these could originate from other platforms or computer systems. This flexibility has not yet been utilized at Headlands Mortgage, yet it does provide them with the means to scale the system's functionality in the future.

The fax robot lives as UNIX user like any other UNIX user. The only difference is that this user is a piece of software that runs in the background accepting mail requests. The fax robot will email back to the originating real user the status of the fax success or failure. Currently the Fax robot has 4 fax modems that are actively being used at one time. The largest and heaviest computer processing is the conversion of the letters into a fax file format. This conversion process can take up to 10 seconds to convert the file to a fax file.

Call Flow

The call flow is one way in this type of scenario. The originators of the fax are staff at Headlands. They initiate a fax request and an email message is generated with the letter. It is sent to the fax robot which analyses the header information in the email message, and then sends out the fax.

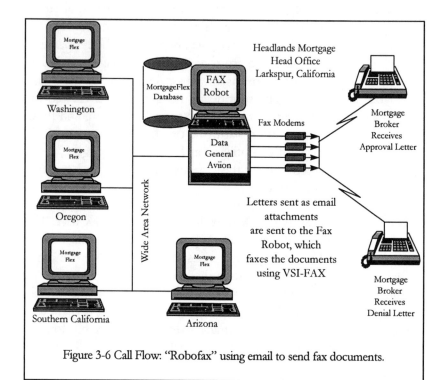

Figure 3-6 Call Flow: "Robofax" using email to send fax documents.

Business Analysis

Headlands has a wide area network from the regional offices to the head office to communicate to a central UNIX server. This was done on purpose to reduce costs and simplify administration. Sending faxes to a central location where one can get the best negotiated price on telecommunications costs, would be an advantageous use of this kind of transmission. Your network costs would be much lower, and your computer and call accounting would be much easier to manage.

If your company has geographically dispersed offices, where one can not justify the cost of providing a private wide area network, the same mechanism can be established using Internet email. Instead of having lower volume higher priced fax traffic at different locations, one could use low cost based Internet email to move the documents to a central location, and have it faxed out at one location.

RETURN ON INVESTMENT

The benefits of having the Fax robot software at Headlands to work with V-Systems VSI-FAX, is evident in the amount of time the system saves in manpower. Given that 300 documents per day are generated, and that on average it would take 5 minutes of manpower to manually fax the document, then this represents 1,500 minutes, or 25 man hours every day, that is not spent across the corporation in trying to send a fax. This is a minimum amount, because this assumes that each fax was successful on its first attempt. This is 25 man hours per business day being spent looking after other customer needs. If one attributed a cost of $10 per hour, one could then attribute a savings of about $2,500 in potential labor costs per business day. This dollar figure represents an efficiency rather than an accrued new bottom line wealth, since Headlands management did not eliminate any positions because of the software's installation.

RIPPLE EFFECT

The mortgage brokers may not directly realize the added efficiency. They see the fax on their fax machine as before, although they may realize that it is being sent a little faster than before.

Vital Statistics

The "Robofax Broker" - Headlands Mortgage	
Cost	*Statistics*
Weeks of labor to install system	1 week writing the Fax Robot in the "C" programming language
Initial System Cost	$6,000
Est. Ongoing support Cost	negligible
System Size	
# Ports - Total/# Fax.	4 Fax Ports
# Users	50 internal users
# Calls/month	6,000 faxes/month
Computer	Data General
Operating System	Aviion - UNIX System V Release 4.0
Industry	*Financial*

The "Robofax Broker" - Headlands Mortgage

Cost	Statistics
Estimated Return on Investment	
Cost Savings	
Manpower equivalency per year	826 mandays per year
Estimated Manpower savings per year	$650,000 equivalent manpower cost per year used to do customer service and corporate tasks. This is not real money saved, but money more wisely spent
Reduction in telephone costs/year	not reported
Earnings	
Generated new income/year	not applicable
Ripple Effect	
Estimated cost Saving to customer base	not reported
Vendor	*V-Systems*
Product	VSI-FAX
End User	Headlands Mortgage Company
Computer Telephony Technologies Deployed	
Fax Server with customized interface	Yes

Table 3-3 The "Robofax Broker" - Headlands Mortgage - Vital Statistics

---

[29] Quote taken from About HMC: Headlands Mortgage Company's World Wide Web Home Page, http://www.headmort.com/abouthmc.html as of 10/19/96 9:43:26 PM

Technical Reasons for Using High Density Fax Servers

In computer telephony applications, one usually associates a more high density fax solution. This is differentiated from the everyday data and fax modem is the hardware technology itself, as well as the kinds of applications demands being placed upon them.

Hardware Considerations

From a hardware standpoint, a more "industrial" system will have fax cards that simultaneously support multiple telephone lines. This can be as small as 2 lines or a multiples of T1 (24 lines) or E1 (32 lines in total of which 30 lines are actually used voice traffic) within the same computer. The fax cards can either be used as a dedicated fax application, or mixed as a pooled resource with other computer telephony hardware such as voice, text-to-speech, and speaker recognition cards that either reside in a PC running UNIX or a UNIX workstations from SUN, or HP, or others.

A high density fax card technology deploys special digital signal processors (DSP) chip sets. Digital signal processors are multipurpose processors, with the appropriate software, which can do many tasks. They can be used for fax, as well as voice, and data, rather than having dedicated products to handle these forms of communication. They are very fast chips that can perform many calculations per second, and in some instances can be as or more powerful as the computer's main central processor.

Fax technology by nature is usually very numerically compute intensive graphical application. A document image has to be has to be converted to a Group III fax format for transmission. Group III fax format is an ITU-T (A United Nations organization) standard. This takes computing power. The standard describes how two fax machines transmit the compressed data of an image. This is done by regular fax machines as the scan the image and transmit the compressed data to the other fax machine for printing. In computer based fax servers the source of the document is in electronic format. This format can be from various sources such as Adobe PostScript, TIFF (Tagged Image File Format), and PCL (HP's Page Command Language popularized by their line of laser printers, and their clones) and text. This can be outputted from various sources, such as image files from a scanner, or stored as a print file from an software application. In that format it is useless to the fax software and the high density fax board. It has to be converted to a form of TIFF file that can be faxed. This is done using either 3rd party or supplied conversion software routines. Essentially a fax file is a form of a TIFF file.

In theory at least, it should be relatively simple to take a fax formatted TIFF file and move from application to application and from platform to platform. However, this is not usually the case, because some vendors of fax software will add proprietary header information to the files, in order to gain better control of the document, and subsequently of their customers. Some manufacturers have tools to convert these special formatted fax files back into uncompressed TIFF files and other popular graphical file formats so that they can be included in other applications as graphic elements. This is important to note that the fax files themselves are graphic elements, and not stored as regular text in a computer. In order to obtain text from a fax document for word processing or text editing manipulation you can:

1. Manually retype the text (most accurate).

2. Use optical recognition software to try to transcribe the image of the words into text.

Because of the nature of the conversion and manipulation of the fax file on a system, the use of high density DSP fax cards is an important advantage over modem based fax solutions. Although modem manufacturers have started to deploy DSP technology to capture voice as well as data and fax for lower density applications, the fax software depends largely for the main computer processor to handle the fax file manipulation and conversion. If thousands of faxes need to be delivered in a matter of a couple hours, then high density fax solution will solve the problem. The effectiveness of this kind of solution is based upon the UNIX operating system as well as the DSP fax technology. Because the UNIX operating system is multitasking, and multi-user, it is very efficient in managing the overall environment than some other operating systems. In a high Input/Output (I/O) bound environment such as faxing, a dedicated fax board off loads the operating system from this I/O task, leaving it to efficiently gather data, manage disk, and memory resources. Fax modems are usually attached by serial port devices, which are managed by the operating system. This places various degrees of burden on the UNIX operating system and the computer system. This is usually alleviated by the use of high end serial I/O cards from Digiboard, or Equinox, or Stallion Technologies. These boards off load the serial I/O tasks, but do not do any of the fax manipulations such as high density DSP boards.

As Alodia Hankins, of T4 Systems, a software developer of BrookTrout high density fax solutions on the SCO OpenServer platform said, that many of their clients have testimonials to the increased speed of throughput as compared to regular fax modem delivery methods. For example, VWR Scientific send 1 million pages of technical safety per year using a T4 MultiLINK system. This saved the company an additional $100,000/year in printing and delivery costs over and above the 4,000 pages/day that are faxed from their call center. The same efficiency and cost savings are played out time and again in different industries such as, health care, transportation, government, manufacturing, and distribution.

Doug Oldham, Vice President Development, of T4 Systems listed the technical advantages of a UNIX based fax server using BrookTrout boards. (Doug's key points are in quotes, with comments by the author are included below.)

1. "More control over the channels" (i.e. fax ports). The fax software allows developers better control over the hardware environment, at a level not usually provided in some software only solutions. Software developers of high-density solutions are provided better programming interfaces.

2. "Richer API for dealing with modems" There is a better application programming interface dealing with high density fax cards, than the programming interface of Hayes modem "AT" command set.

3. "Better response from internal hardware": Since the high-density fax cards are on the high speed computer bus, and they are in turn are communicating on a high speed MVIP, PEB, or SCbus telephony buses to the outside world, there is greater throughput to the digital or analog telephone lines. Data utilization and transmission are at the best possible speeds.

4. "Reduced cables and peripheral components neater packaging." A problem with high density rackmountable modem solutions, there is more cable management, than just terminating a single digital T1 line to a T1 card inside the UNIX system chassis. Cable management may seem trivial. If you have a PC on your desk just look at the cables needed to connect a phone and modem on a single analog telephone line to the outside world. Now think about attaching 24 or even 48 serial cables from the back of a UNIX system to a rackmount of fax modems. It takes management time to document their connection, and make sure the correct serial port on the computer is connected to the correct telephone jack coming out of the wall or punch block. It can be done, however, just connecting a single T1 cable makes a much neater package. Besides, in most Canada and US tariffs it is more economical to use a digital line for such high densities.

5. "Downloadable firmware for updating without replacing hardware:" Firmware provides the software intelligence for a hardware device. This is different than the application software a user uses to fax documents, or execute other programs. Such firmware can be loaded or "burned in" to chip that is physically mounted on the circuit board of the device. This means that the computer system has be turned off, remove the boards, and replace the firmware chip. With products such as BrookTrout's this is simply an set of software instructions that can be downloaded to the remote high density fax server, and update this firmware. This is an important feature, when a system is sending out many thousands of faxes a day. Only some of the top of the line modems have this feature.

6. "IVR (voice) features": This is further discussed below. It means you can add more additional value such as IVR, text-to-

speech and other technologies. Lower density fax modem solutions have limited voice capabilities for basic voice mail play and record.

7. "Expandability is easier" This is particularly true, since adding greater increments of high density fax board solutions is simply adding another board, and software. If the system has used a "Fractional T1" (i.e. the original installation had used only 8, or 12, or 16 ports of the T1). An additional fax board can complete the T1, or simply expand beyond it with more T1s. There are no additional computers, or other major components. It is one contained system.

Another advantage from a hardware prospective is that with a higher density, it is more economical to use digital T1 or E1 lines than it is to use a large number of analog telephone lines. One reason is that if one wanted to send and receive from 24 or 32 fax digital lines, then it would be easy to install a single T1 or E1 card from Acculab or Dialogic, or another vendor, and then connect the fax boards using a MVIP or SCbus cable to the fax boards. A single digital line is put into the system, and then the fax software has access to all the inbound and outbound traffic. This is not unique to a UNIX configuration, rather more of a statement of how it is done. However, if one wanted to have 96 lines of fax within a single system, then the UNIX based solutions will shine through as true champions. The design of the operating system will make a big difference in managing resources. Fax cards come in various densities from as small as 1 port to 32 ports on a card. In a PC UNIX environment rack mountable industrial PCs that have 20 ISA slots that when the basic components are installed, can have up to 15 slots available to place fax, and digital interface cards.

Besides the density, another major benefit is the ability to mix and match other types of voice processing boards to create a more complex and multimedia solutions. In this environment, the fax boards act as a pooled resource either in as a smaller ratio to the number of voice cards, or if application requirements demand it, in a one to one combination. Depending on the manufacturer, it is possible to achieve this by adding more fax boards within the chassis of the system, or in the case of some manufacturers, such as Linkon Corporation, a software license can be issued to increase the number of available fax ports on an existing card running inside the system. This is a clean solution since the system administrator locally, or even remotely can increase the fax resource without having to make a trip to the site, and open the cabinet and have an extended downtime. This again is a testimonial to DSP based technology.

Another advantage that a UNIX system has with high-end computer based faxing is its ability to easily internetwork with other networks or mainframe data sources. Allowing applications from different sources to drive a very large volume of faxes to a UNIX based fax server.

There are many hardware suppliers of fax based available in products from various companies such as Dialogic's Gammalink product line, BrookTrout, and Linkon Corporation only to name a few.

Application Considerations
In the high-end computer based fax market, the need for stable, dependable and large volume systems is evident in the kinds of applications that are used with this technology. These include[30]:

1. "LAN Fax: This technology allows network users to share fax hardware resources and send and receive faxes directly from individual workstations." This is very much the same role that is described in the case histories within this chapter. The difference is that the hardware configuration uses higher-density solutions.

2. "Image Servers. These give users the ability to input, store, retrieve, and process information. Companies that rely heavily on paper-based information to make business operations can be more efficient." This allows for paper documents to be faxed to a central location, and then redistributed by schedule or on demand to other fax locations. This centralizes the paper documents information, allowing others to access it, especially if they can manage it using a database, and IVR components. An potential user of this could be a customs broker, which would accept customs documentation by fax from a employee

of a transportation company (i.e. truck driver, or shipper/receiver) and then fax it to a central fax system, so that all parties involved (the brokers, sales team for the broker's customer, and others) can get copies of the documentation.

3. "Broadcast Servers: These centralized systems are for unattended faxing of high volume and batch faxing, often during off-hours". These are used to send faxes in great volume at the least amount of time, at the least cost. Product announcements, press releases or technical information, and as well as unsolicited "junk fax" are example applications of this kind of high density fax solutions. The fax numbers and the fax information are driven from a customer database.

4. "Host Fax. These servers are used to fax high-value, high-priority documents or transactions". Possible applications would be in the financial industry, where network servers, or mini or mainframe host computers can directly output large number of faxes in a very short time.

5. "Fax-on-Demand." This combines components of IVR with fax services. Callers can request fax documents 24 hours a day 7 days a week using their touch-tone phone, by requesting a document by number in response to a voice prompt. The document is then scheduled to be faxed to a destination fax number supplied by the caller. The fax delivery can be either one-call or two-call methods. The one-call method requires the caller to call from a fax machine's telephone handset. They call the service, and request the documents they want. If they do not know the document number, they can request a catalog of documents and numbers, and retrieve the documents they want on a subsequent call. Once they have finished selecting documents, they are asked to press the start key of their fax machine. While in the voice prompt and touch-tone collection mode, the caller is interacting with the voice card component of the system. (Either this be a dedicated voice board linked by a telephony bus cable to a fax resource card, or an all-in-one DSP based card that has both voice functionality with fax software on the same card. Example manufacturers of these

types of cards are Linkon, and Dialogic). Then the caller is told to press the start key on their fax machine, and the fax card then is engaged and the fax machine prints out the requested documents.

A two-call scenario is where the caller requests the documents, supplies a destination fax number, and hangs up. Then the fax card sends the document to the fax number supplied by the caller. The two-call method is the most convenient. The one-call is the least costly for the owner of the fax-on-demand system, since all long distance toll charges are incurred by the caller. CallStream Communications with their FaxStream module of their VoiceStream IVR product provides both kinds of solutions on the SCO UNIX platform. Other UNIX developers such as Apex, and Mediasoft do as well. An important note is that fax-on-demand applications still provide an excellent means of customer service and support.

6. "Service Bureaus. These provide services such as never-busy mailboxes, store-and-forward, extremely high-volume broadcast, and least-cost routing to businesses".

Another application issue is the fact that the Group III standard does not necessarily support an easy mechanism to personally deliver fax to an individual. This is a problem if confidential legal or financial information is laying out for everyone to see in the fax machine pool. One method of solving this is using Direct Inward Dialing where a distinct phone number is published for a particular individual. Instead of this number being terminated on a private fax machine the faxes are delivered into a fax server. The DNIS on the line is captured, giving the number that was called, and the fax is accepted and stored for the individual with that phone number. This then allows for the fax to be retrieved by the individual using an fax-on-demand type of delivery, that is either stand alone or part of an integrated messaging voice mail system. This method is popular in North America, while in Europe with the proliferation of ISDN technology this is done a bit differently. An ISDN line allows for more than one distinct phone line to be associated on the same cable.

These and other application demands have driven computer based fax market from $28 million per year in 1988 to a projected $1.7 billion by the end of 1996[31]. Not every company and individual in the world has or can afford to have access to the Internet and the world wide web. Documents provided by fax can be delivered up to any of the 10 million or more fax machines in the world today. One UNIX based fax-on-demand product allowed customers to call a number and then "surf" Internet world wide web pages. This allows for a common document source pool from which both Internet users, and fax-on-demand users can draw upon.

Both low density, and high-density fax server products on UNIX provide an important resource in reducing company costs, and improve service for their customers. As the convergence of communications sources continue, fax has become more blended into other applications. This has been already witnessed in the realm of integrated messaging, where fax along with email, and voice mail are presented on a single system acting as a client to a centralized or even distributed set of servers. Internet based faxing of documents are available using browsers, such as performed by Faximum Software. Fax server technology on UNIX will continue to be an enhancement and integral part of overall communications technology.

---

[30] "The Computer Based Fax Market",
   http://www.dialogic.com/products/gamma/market.html, as posted on the Internet world wide web page as of January 21, 1997 at 11:42 PM. Copyright 1997 by Dialogic Corporation. Those items in quotes are from the posted article. The remaining text is additional material by this author.

[31]The Computer Based Fax Market",
   http://www.dialogic.com/products/gamma/market.html, as posted on the Internet world wide web page as of January 21, 1997 at 11:42 PM. Copyright 1997 by Dialogic Corporation.

Computer Telephony Integration - Call Centers

omputer telephony integration is the specific term used to describe the coordination and delivery of both corporate data, and live voice data to a person within an organization. This has been achieved through proprietary interfaces and products provided by large telephone PBX manufacturers for years. With the advent of newer lower cost technologies such as TSAPI, TAPI, and CSTA these are now moving out of the traditional model, to more network, and lower cost delivery of these important services, to smaller workgroups within a company.

The following is a review of how computer telephony integration at very large site saved hundreds of thousands of dollars per year, and helped increase sales and managed credit. The other case history reviews Teloquent, an SCO developer, which uses ISDN communications technology to provide a decentralized "virtual" call center environment for Georgia Power. A technical discussion ensues, reviewing the kinds of private branch exchange and data integration strategies that allow for computer telephony integration on the UNIX platform.

The "Retail Edge" - Federated Department Stores

Figure 4-1: Federated Department Stores: Using IBM's DirectTalk & CallPath products Federated has increased sales, while millions of customers have saved time.

What would you do if you had to answer tens of millions of calls per year? Look for a better way to cut down costs and increase efficiency! In an environment of this magnitude, finding even a small time saving means dramatic bottom line savings in manpower and telecommunications costs. In an era of slack consumer demand, cost savings mean better return on investment for shareholders. Federated Department Stores Inc. use of IBM's DirecTalk & CallPath products in this call center environment truly demonstrates the magnitude of cost savings that computer telephony integration promises.

Background

"Federated (Federated Department Stores Inc.), with corporate offices in Cincinnati and New York, is one of the nation's leading department store retailers, with annual sales of more than $15 billion. Federated currently operates more than 400 department stores and 150 specialty stores in 36 states. Federated's department stores operate nationally under the names of Bloomingdale's, The Bon Marche, Burdines, Goldsmith's, Lazarus, Macy's, Rich's and Stern's.[32]"

Don Lykins[33], Group Manager Systems Development, is responsible for managing the call center in Cincinnati, Ohio for the FACS division. Federated is managed in different divisions. There are eight operating retail divisions, and three supporting divisions. These are Federated Merchandising/Product Development in New

York, New York, Federated Systems Group in Atlanta, Georgia and Financial and Credit Services Group (FACS Group), in Cincinnati, Ohio. Each division is autonomous. Each division has a chairman and president, who report to Federated CEO Allen Questrom and President James M. Zimmermam. Seven years ago there was no FACS group. Each department store had their own computer and credit card authorization systems. In an effort to consolidate costs, the FACS group was created to provide a single source of credit information and management and customer service.

Glossary Alert!

➤1 terabyte is 1 trillion bytes of data or *1,024,000,00 0,000 characters.*

The FACS group is based in Mason, a suburb of Cincinnati. Federated has a large commitment to proprietary credit card management for the stores. The FACS division provides the credit card services for most of Federated department chains with tens of millions of credit card customers. The size of the operation is very impressive. The call center occupies 5080 square meters (200,000 square feet) of physical space, which can house up to 1,500 telephone agents at one time. The AT&T G3 telephone switches handle that load and can support up to 2,300 calls at one time.

System Overview

In order to manage the terabytes size database systems for this huge volume of customers, the mainframe systems are managed by the Federated Systems Group. The corporate database systems in Atlanta are based upon Datacom, using IBM host mainframe equipment, while the FACS division uses a client-server architecture with Microsoft SQL Server with IBM host access to the Datacom databases.

In an era of client-server computing, the need to move the data from the mainframe to the new data sharing model was found to be not as advantageous as it may have seemed at first. Because of the large amount of software that is specifically customized for the needs of Federated, and the large transaction volumes the mainframe can handle, the mainframe systems have become the central repository of data for Federated. In this environment, Don's applications gather the data from the IBM host screens, collects it and reorganizes it for presentation to the customer service representatives using client-server technology.

FACS has three call centers throughout the country, namely Phoenix, Arizona, Tampa, Florida, and Mason, Ohio. About half of the call volume is from customers calling from home to obtain the balance of their credit card account, and to update their address

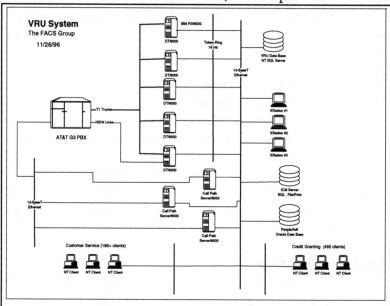

Figure 4-2 The Voice Response Unit System for the FACS Group. Copyright Federated Department Stores 1996. Used with special permission.

information. Those calls go to the customer service department, to any of the three call centers, depending upon the Federated retail division for which the caller has a credit card. It is possible for a customer to have more than one credit card from the different retail divisions of Federated, yet all the customer service calls go to the same customer service department at the different call centers. The other half of the calls are for opening new accounts, lost or stolen cards, and credit limit approvals at the point of sale.

Structurally, the call centers each have AT&T G3 private branch exchanges, which communicate with up to five IBM DirecTalk/6000 Server systems. The DirecTalk/6000 systems also communicate with the mainframe systems in Atlanta, Georgia. Depending upon the location of the call center, the number of interactive voice response telephone ports range from 360 to 504. There is currently 1344 telephone ports in total.

Call Flow

In the scenario where customers in the northeastern United States wish to check their account balance, the caller would use a toll free 800 number and the call would arrive at the main call center in Cincinnati. The call is answered by an IBM DirecTalk/6000 interactive voice response system, running on an IBM RS/6000 platform using AIX, IBM's implementation of the UNIX operating

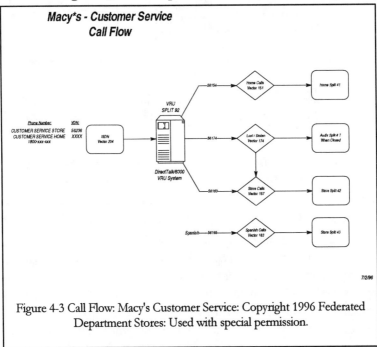

Figure 4-3 Call Flow: Macy's Customer Service: Copyright 1996 Federated Department Stores: Used with special permission.

system. The caller is then prompted to enter their account number. The inquiry is made, and the data is collected from the mainframe system, and the caller is provided with his balance information.

If the caller also wants to inform a live agent of a change of address, he presses "0". IBM CallPath/6000 product takes the

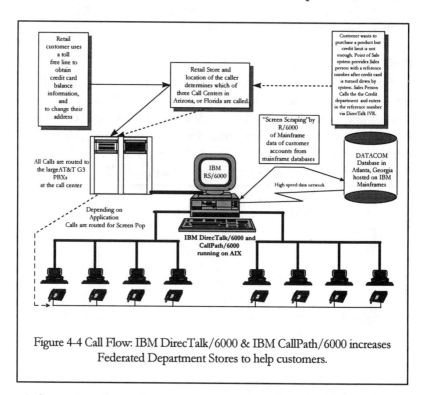

Figure 4-4 Call Flow: IBM DirecTalk/6000 & IBM CallPath/6000 increases
Federated Department Stores to help customers.

information about the customer, communicates with the AT&T G3 switch, and locates an unoccupied customer service agent. CallPath instructs the AT&T G3 to transfer the call, and then transfers the customers data to the WindowsNT system at the agent's desk. The agent has the "screen pop" to their workstation with live telephone audio and current client information. It is important to note this eliminates the customer having to tell the customer service agent any new information about who is calling. This process is fast and accurate and is the key to the understanding of how the elimination of that time means a reduction in costs.

Another major application implemented by FACS has been the integration of the point of sale credit authorization. This is one of the largest applications that Don has worked on in his career. A customer wishes to purchase a $2,000 item. The customer goes to the sales person at a cashier counter which has a fully integrated point of sale system. The sales person swipes the credit card through the credit card reader. Unfortunately, the card holder has only $1,500 available credit.

Before computer telephony integration had been put in place for this automation, the sales person, would call the call center and talk to a customer service agent, who may authorize the purchase. This would take some time as the sales person would have to wait in the call queue, before accessing an agent. Today, the point of sale system notifies the sales person and produces a unique code on the point of sale terminal. The sales person then calls the credit service department, and a DirecTalk/6000 interactive voice response application prompts for the special point of sale terminal code. This code is used to access the pertinent credit information about this customer. CallPath/6000 finds the next available agent and forwards this information to the screen for review. It shows what the customer is trying to purchase, and the credit limit. The customer service agent can approve or deny authorization of the purchase, based upon established rules and procedures.

In either case, the call is transferred back to the CallPath/6000 system to announce the results to the sales person. If the purchase can be made the transaction is processed. The key thing is that the input, human decision making, and the results are completely automated, resulting in a much quicker turn around for an answer for the customer. It also reduces manpower costs at the call center. This shows how one application, such as point of sale, can be integrated with another application outcome using computer telephone integration. It demonstrates how to save money because it saves time "on the front end as well as the back end" of the process.

Business Analysis

The evaluation of vendor solutions was part of the cost justification process. After reviewing a TSAPI versus the IBM CallPath solution in late 1994, Don Lykins decided that IBM provided the level of capability they needed. They implemented a prototype system for a 2 month period for all call handling scenarios, with 8 agents. The results from the prototype demonstrated, the "screen pop" technology could reduce by 20% the talk time it took a customer service agent to handle the call. A lot of the talk time was reduced by eliminating the need to have the customer speak the account number to a live agent. The interactive voice response application captures the required information and forwards it through the system to the appropriate agent. In hindsight, Don Lykins said "the IBM solution was the best solution" after they had learned the issues they faced.

RETURN ON INVESTMENT

The 20% reduction in terms of agent time cost, and 800 toll free minutes translated into "huge savings" for Federated. This applies for both the customer balance inquiry, with change of address application and as well the point of sale system. Individually these applications have averaged about 20% reduction of the call handling time.

The customer service department has about 200 workstations that over the different shifts have about 300 agents working from 8:00 AM to 11:00 PM.

They have met their goals of providing the same service levels with a reduction in staff. In general terms, the cost reduction objective has been the amount of staff working in the call centers. The IBM systems in the Ohio site alone have had a net effect of providing the equivalent workload of over 180,000 calls per month. Hence there has been a major impact of between 10% to 15% reduction in staffing requirements for each of the computer telephony applications implemented so far.

RIPPLE EFFECT

For the customer, it has saved the time they needed to enter their account number. Before the computer telephony integration, the customer had to enter the account number once if they used the older interactive voice response system, and then had to say it all over again to a customer service agent. Since the account numbers range from 9 to 16 digits, depending on which credit card they have, this added to a significant delay in service, as well as an irritant to the customer. If one takes the estimated savings of 180,000 calls per month, and assumed that to recite an account number to a customer agent takes about 10 seconds on average, then this represents about 62 man days each month (1 manday is equal to 8 hours) in savings as averaged over all the customer calls. With the new implementation Don said, "Customers find it a great time saver since they now only have to enter the account number once."

Vital Statistics

The "Retail Edge" - Federated Department Stores

Cost	Statistics
Weeks of labor to install system	16 weeks
Initial System Cost	$731,000
Est. Ongoing support Cost	$15,000
System Size	
# Ports - Total/# Fax.	300 to 504 ports depending on location.
#Users	millions
# Calls/month (Inbound Voice Response Units)	713,000 (Customer Service Only)
# Calls/month	475,000 (Agent)
Computer	IBM RS/6000
Operating System	AIX
Industry	*Retail*
Estimated Return on Investment	20% reduction in call handling time
Cost Savings	
Manpower equivalency per year	
Estimated Manpower savings per year	10% to 15% reduction
Reduction in telephone costs/year	not reported
Earnings	
Generated new income/year	not applicable
Ripple Effect	
Estimated cost Saving to customer base	62 man days each month
Vendor	*IBM*
Product	DirecTalk/6000 & CallPath/6000

The "Retail Edge" - Federated Department Stores

Cost	Statistics
End User	Federated Department Stores
Computer Telephony Technologies Deployed	
Interactive Voice Response	Yes
Computer Telephony Integrated "Screen Pop"	Yes

Table 4-1 The "Retail Edge" - Federated Department Stores - Vital Statistics

---

[32] Quote taken from Federated Department Stores world wide web site on the Internet, on 09/21/96 1:54:51 PM, 2nd Quarter 1996 Results. http://www.federated-fds.com/

[33] The author wants to thank Don Lykins Federated Department Stores FACS Group for going the extra distance in providing insightful input, and access to call flow, and system diagrams, cost and call volume information for this case history.

The "Virtual Distributed Telecommuter" - Georgia Power[34]

How do you maintain cost effective satellite offices to your existing call centers without having dramatic incremental costs to deliver software, hardware and manpower at those locations? If your company or subsidiary is United States owned, how do you help your company meet the requirements of the US government Clean Air Act? This requires certain corporations of specific size to implement telecommuting to reduce ozone emissions caused by staff commuting with their cars. The answer is virtual distributed call center technology. There are definite hard cost savings in using this kind of technology. Of interest in this case history is the increased human wealth that the company gains, as a result of forging strong employee company loyalty and provide the means to increase employee's life style and family relationships.

Background

Southern Company is the holding company of Georgia Power, Gulf Power, Alabama Power, Mississippi Power, and Savannah Electric. Southern Company is the 2nd largest investor owned utility in the United States. Georgia Power provides the generation and distribution of electric power throughout the state of Georgia.

Originally Georgia Power had nearly 300 different published phone numbers. These would terminate at the local offices of Georgia Power throughout the state. As the number of customers increased, this did not always provide the level of service expected by the public. In order to provide better service Georgia Power built two call centers that were linked together. One in Macon and another in Atlanta. Although Georgia Power is migrating to client-server architecture, the call centers' main database application is based upon CICS technology from IBM. The PBX environment for both call centers is Teknekron. In order to simplify customer service, Georgia Power forwarded calls from the 300 individually published numbers to a main call center number, to handle questions such as opening new accounts, high bill complaints, general service questions, and to report power outages. With the increase of telephone traffic, the interflow of the two different call centers compounded the complexity and management issues. Adding more to the existing systems was not the answer. This would only increase problems at the existing call centers.

The call flow from the remote locations to the call centers was not very efficient.. There was no computer telephony integration at the time, and they did not have the software release levels in the ACD of the Teknekron PBX to perform the screen pop functions. Major upgrades to the ACD would be needed to provide this functionality

Another motivation to use virtual distributed telecommuting technology for Georgia Power was the need to comply to the Clean Air Act. In 1992, the Atlanta chamber of commerce asked Georgia Power to look into a telecommuting program. As part of the effort by the chamber of commerce, it was recognized that there were two ways to help reducing commuter car pollution. The first is the use of public transportation, car pooling and other methods which are employee decisions. The other is telecommuting which is manager decision driven. This is largely driven by information resources and telecommunications capabilities. The success at Georgia Power was used as a model for other companies to use.

Cars are responsible for 70% of low level ozone emissions, which greatly contributes to smog in urban areas. Atlanta is considered to have a low level ozone emission at the serious condition by the Environment Protection Agency of the United States government, while Los Angeles, California is considered to have an extreme level. There are different sanctions and requirements at each of these levels. Atlanta has developed a program to come under those guidelines. However, in the spring of 1996 ozone emissions were getting worse, because of the growth of the city especially in the outlying areas. The 1996 Summer Olympic Games experience showed that the city was able to obtain a 30% reduction in rush hour traffic. During the 17 day period of the Olympics, the city of Atlanta was not out of compliance with the Clean Air Act. This showed that it was possible to reduce emissions.

Not every job is a candidate for telecommuting, nor is every staff member a candidate for telecommuting. One survey Frank Boyd referred to was that 30-35% of a business personnel are candidates for telecommuting. In manufacturing environments this average will be lower, while in insurance the average will be higher. Not every one can work independently. However this is not the case at the remote call center.

Generally, when reviewing the kinds of employees that are candidates for telecommuting, it is advisable to look at both hourly and salaried employees, and the kinds of work they were doing. Hourly employee positions that are candidates for telecommuting are:

1. claims processing

2. employees involved with handling accounts payable and receivable.

3. purchasing

4. customer services using the phone, where you do not have to be at a physical location and where you do not have access to physical things other than a computer and a phone.

On the salaried sided, those who work with customers and are dealing with the outside public on an ongoing basis are also candidates for telecommuting.

There were several reasons why Georgia Power was looking to expand their existing call center capabilities.

1. The number of customers in Atlanta, Georgia increased, and the size of the toll free calling area was also growing. This may be seen as a costly new business requirement but it also has its advantages. The expansion of the toll free area allowed for the redirecting of calls away from some smaller call centers outside of Atlanta.

2. The size of the call center in Atlanta Georgia was 100 agents. The increased call handling capacity requirements placed demands on the physical space at the call center. They were running out of space.

3. The Clean Air Act started forcing Georgia Power to change the commuting patterns of its workforce. Georgia Power needed to meet the goals of obtaining a number of credits provided by the Clean Air Act. In order to meet these goals they needed to have less staff commuting to their call center and for other workers to their offices. In this case the commute was into Atlanta, Georgia. In general the goal of the Clean Air Act is to have up to 20% of workforce telecommuting for specific sized businesses by the year 2000.

Georgia Power deploys smaller remote call centers that are an extension to the existing call centers. The communications between the main call center and the remote call center is done by using dedicated T1 connections and a remote shelf off of the ACD. (A shelf is a term used to describe a remote call center office that is like an "extension" to the existing call center.) The data communications component also requires a dedicated T1. The net result is that a remote shelf requires a minimum of 20 staff members to be cost effective. Even if they used a fractional T1, the other investments in hardware and labor to set up the shelf did not make it practicable .

Another factor in seeking a newer configuration and solution, was that the increased demand also meant increased investment at the existing call center. The investments were in the physical space and computer and telephony technology. There was a capacity of a maximum of 30 locations on the computer host side. Using remote shelves everywhere would have led to sites in the southeastern, southwestern, northeastern, and northwestern portions of the city. "This was not practical.", said Jerry Cohen, of Georgia Power. Networking more independent call centers is even more costly, because of the need for separate databases, and reporting functions at each location.

System Overview

Georgia Power was determined to define the problem and design a solution first before looking at products to solve the problem. They wanted to ensure that the solution was not biased by or key holed into a specific product offering.

IDEAL SOLUTION DEFINED

Jerry Cohen stated the ideal solution was, "To separate the call routing and the intelligence from the ACD, (automatic call distribution system), centralize the call routing along with the intelligence while distributing the agents wherever we wanted to. We were aware that this was possible but we were having trouble identifying traditional solutions providers who could do this. From the beginning, the key was our local exchange carrier (Bell South). We needed them to provide the distributed connectivity. At that time they were forming a business relationship with Teloquent." Because Georgia Power was able to articulate what the solution was, Bell South was able to work with them to implement Teloquent.

Call Flow

The initial implementation of the Teloquent's Distributed Call Center server was at the Atlanta call center, from which Georgia Power communicated with a remote office in northeast Atlanta with 12 agents. At the beginning of the project, Georgia Power had all the calls terminate to the legacy Teknekron PBX. The Teknekron would offer calls to the Teloquent system. There was no intelligence passed between the two systems and, as a result, they were unable to take advantage of many Teloquent features. The Teknekron PBX dumped the call overflow and other calls into the Teloquent system to be answered by the remote 12 agents in northeast Atlanta. Jerry Cohen pointed out that the calls did not necessarily have to be placed to that remote office. Initially the CICS database application information was not screen popped. The reason for that was, the Teloquent system was residing behind the Teknekron PBX. The Teknekron did not pass onto the Teloquent system ANI (caller-id) information that would allow for a screen pop to be delivered. The first phase of this project created what was a more effective remote shelf in the traditional sense of the legacy PBX technology. It provided greater flexibility in the fact that management could now have the agents anywhere they wanted. This was possible because of the flexibility of the ISDN support and Teloquent's Distributed Call Center design. This did not require the same number of agents as did the remote shelf technology to be cost justifiable.

Another example of the flexibility was the ability to move the limited number of agents on the late night "graveyard" shift. Georgia Power realized they could save on supervisor labor costs by moving these "graveyard" shift workers in with other staff on the same shift at Georgia Power's distribution center. By making this move, a single supervisor could manage the 4 customer service agents, and the 4 distribution monitoring staff at the distribution center.

This provided an opportunity to reverse the Teknekron and Teloquent configuration, placing Teloquent's Distributed Call Center system to front-end all the calls. Now the Distributed Call Center played the role of the master handling system for the entire call center. The calls would go from the Teloquent to the Teknekron. The Teknekron was still used to support the customer service agents in the local call center. The Distributed Call Center received caller-id and now could perform screen pop capabilities for the agents in the northeast satellite office. The CICS database application screen pop integration occurs through transaction processing at the workstation level, rather than at the mainframe level. This means that once the call arrived and was passed onto the next available agent in the remote northeast call center, the screen pop was made via the workstation issuing the CICS application command, rather than attempting to directly command the IBM mainframe to spawn the database transaction, and to send the screen of customer information.

One of the side issues that showed the difference between the older CICS/Teknekron system versus the newer CICS/Teloquent system was the productivity gains by placing the phone function as software inside the agent's PC that had a Teloquent ISDN card with Teloquent phone software. The major differences were threefold :

1. Those who had migrated from using dedicated IBM terminals used for the CICS applications, and later migrated to PCs running IBM mainframe terminal emulation software, did not migrate as easily as did those agents that had the Teloquent system on PCs. Those agents that had only the PC with the Teknekron delivering the call and then performed the CICS transaction using regular key entry were not as happy, nor were they as productive.

2. Those agents that had the embedded Teloquent phone technology within their PC were more efficient in the combination of the phone presentation, as well as with the CICS data.

3. By placing the phone "inside" the PC, CTI applications such as screen pop were very easy to implement. As a further benefit, additional investments in software or hardware were not needed.

As for ongoing support and system administration, Jerry Cohen said the Teloquent GUI interface used to manage the call routing and other system administration tools, provided an easy to use and friendly environment to manage the system. This was a benefit because the call center was not traditionally a UNIX user, and they found that they could use existing skill sets based upon PC GUI technology to manage the system. Hence, they did not have to invest heavily into training of system administration staff in the UNIX environment, in order to provide ongoing system maintenance. Since they were using a PC based UNIX solution from SCO, Georgia Power was able to take advantage of their existing purchasing contracts with Hewlett-Packard for equipment and hardware support.

At its peak, 28 customer service agents were being served by the Teloquent system in a distributed virtual call center environment. Initial implementation took about 6 weeks to install the Teloquent behind the Teknekron. Some of the factors were that the integration of the Teknekron extended the integration time. When the Teloquent was changed to take all the calls, then it was much easier and quicker, since they had to only work with the local telephone company to redirect the calls.

Teloquent was also used in a pilot project of 5 agents working from their homes. This reduced the cost of the physical space demands. Since the cost of an ISDN line in Atlanta is only $60 per month, this pilot project helped test the feasibility and flexibility of virtual distributed call center technology in this operational mode

Georgia Power has over 100 people currently doing telecommuting. 72 staff members use a product called Reachout to communicate with the Georgia Power's LAN and use regular analog telephone services which are not integrated into a computer telephony integration environment. As the ability to deliver more ISDN services to the Atlanta area develops, it is foreseen that the Teloquent system will be used for other services. It is forecasted that an additional 45 customer service representatives will be working out of their homes, using the Teloquent system, if the pilot program 5 agents is successful.

Business Analysis

Although the business case for telecommuting and a virtual call center is strong both in terms of governmental, and economic factors, there are benefits derived that are considered "soft" human benefits, that result in better performance and employee loyalty. Although it was not directly stated by Georgia Power managers, there are implicit and real ripple effects for Georgia Power's customers and as well as for the citizens of the state of Georgia as a whole.

RETURN ON INVESTMENT

The benefit was the award of the credits from the Clean Air Act which were not elaborated by Georgia Power managers interviewed for this case history. However, this is an economic gain

that has had bottom line impact, as well as it has been a strong
public relations asset to the company.

Hard cost savings for Georgia Power can be measured in
comparison to the remote shelf technology, versus the Teloquent
system.

The cost savings of not having to provide the T1 and the other
sundry equipment to support it, reduced the cost of putting a
smaller size call center together. "In the traditional method, I could
add all the agents I wanted, but unless I added trunks for the agents
I would not increase my capacity. On the other hand, with the
Teloquent system, every time I add an agent I add a corresponding
trunk," said Jerry. The reason is that the ISDN lines terminate at
the central office, not inside the call center on dedicated ACD
ports. "So for a company like an utility that has these very large
peaks, spikes really, in terms of call handling capacity are virtually
unblocked, since all the agents for all the calls can be on the phone
at once. If I had all the agents logged in at once at the call center
with just using the traditional Teknekron model, then half of the
agents would not be active since there would not be enough trunk
lines. One could offer more trunks, but they decided not to share
as many workstations, which is typical." Jerry said that until the
Teknekron has been completely eliminated, then there is not a clear
hard saving that can be attributed in terms of percentages, in
difference in costs between the two systems. However, Jerry did
say that it is about 20% less expensive to deploy the Teloquent
solution as compared to the remote shelf architecture. Jerry
estimates with other vendors such as AT&T and others, that on
average it is about $3,500 per agent using the remote shelf model.
This means a Teloquent agent costs about $2,800 to set up for
Georgia Power. For the 28 agents this translates into about $19,600
in initial setup costs. Other savings in comparison to the
Teknekron system are not as defined for further discussion.

One of the advantages of the Teloquent system is that each agent
has the exact same system no matter where they are. Jerry calls this
"no compromise telecommuting". Workstation setup in the field is
exactly the same as in the call center. This is not the case with other

traditional call center vendors, where there is always some kind of sacrifice in audio quality and/or ACD functionality to accommodate the remote locations. With the Teloquent system and the consistency of the workstation environment, this has reduced some of the support costs associated with these systems.

Another requirement was the need to increase the call handling capacity through automated services, supported via voice response units (VRU). VRU is a synonym for an interactive voice response (IVR) system. The voice response units were originally designed to be stand alone. With the increase call volume demands and with the Summer 1996 Olympics in Atlanta, the need to use VRU's for call routing, and linking them with the Teloquent system's voice service module, was a key component in helping to route callers to a IVR solution or to a live agent. It was a very cost effective method of call handling. This cost 80% less than using traditional VRU methods. "The beauty of the VRU integration was that the VRU units were installed in a third remote location to be located closer to the mainframe for easier support," said Jerry.

Georgia Power was able to save $100,000 in leasing costs, when they were able to put their area development team of 15 employees onto the generic telecommuting program. Although this particular saving has no relation to computer telephony applications, it showed management that telecommuting had bottom line results

Both Jerry Cohen, and Frank Boyd, a manager responsible for telecommuting at Georgia Power, discussed the employee relationship issues that were seen as benefits and return on investments in the use of the Teloquent system.

Jerry noted that one of the benefits of ISDN utilization was the ability to call on a 2nd line. When the agent's children came home from school around 15:00 hours each school day the agents would receive a call. They did not have to answer the call, since they would recognize the caller-id (ANI) on the line and would know that the children have arrived home. This was seen by the agents as being a major benefit.

Frank Boyd commented that employee productivity went up 10% to 15% very quickly. The reasons for the productivity increase were:

I.   Georgia Power carefully picked their better candidates for telecommuting.

II.   The speed of the ISDN and Teloquent phone presentation increased the call handling capacity and was able to provide better throughput to the agents.

III.   The ability of the agents to be close to home. The average commute distance in Atlanta is 45km (28 miles). Frank Boyd indicated that some of the employees take great comfort in that they do not have to face the traffic. It is a powerful morale booster. On average, the employees are gaining 1.5 hours of time per day. This makes the employees extremely motivated and dedicated to the telecommuting program. Employees have commented that they use the 90 minutes per business day to attend to their families, or use the time for personal fitness to work out at the local fitness club. For some employees who have school age children, they do not have to leave for work as early as they use to. Now they leave about the same time their children need to go to school. Frank Boyd commented that "One employee said (to Frank) that (telecommuting) has allowed her to get her life back."

Furthermore, Mr. Boyd said "there are few things in life that you save money, gain productivity, where employee satisfaction goes sky high, and where you are closely aligned to the employee's satisfaction. Telecommuting employees see this work style as a major benefit. This keeps employee loyalty very high". As an indication of the employee satisfaction, Frank said on one visit to the remote call center an employee leaned out of her cubicle and gave him a "thumbs up sign" as he walked past.

Another factor was the reduced down time of the systems in the new configuration, which increased employee morale.

RIPPLE EFFECT

There are two ripple effects of note:

1. Reduced car emissions - better for all involved in the community as well as for Georgia Power economically. The reduction of 28 commuters, as compared to the hundreds of thousands of commuters in the Atlanta area is almost not measurable in terms of emissions or other soft costs to the community. However, it is a beginning. The ripple effect here is not directly to Georgia Power's customers, but rather to the community as a whole. It is a beginning that is now a true reality that can provide better service for customers and to communities as well.

2. Given that the Teloquent system has allowed for better integration for CTI and screen pop, then it can be *assumed* that for the 3.5 million calls per year that the Atlanta call center and its staff are concerned with, this may result in saving an average of just 5 seconds or more per call. Since 28 users are now receiving better screen pop out of 100 agents, then assume that 28% of the 3.5 million calls per year at the call center are being more efficiently handled. This results in about 170 mandays saving per year for every 5 seconds shaved off the call duration. This time is saved by both Georgia Power in terms of manpower and as well as for their customers.

Summary

Overall, this case history shows how economical and practical telecommuting is. It shows the power of using digital technologies such as ISDN in order to provide both voice and data delivery. It also reflects a change in life style and corporate culture that is happening today, because of the capabilities of computer telephony integration technology. Virtual distributed telecommuting is now a reality, and not just fiction. Many companies can realize economic as well as human benefits. It is a reality because market forces are at work, rather than because some governmental edict has forced business to change.

Although prophesied before by many analysts that computer technology will fundamentally change the venue and method of work, the truth is that it now makes even more pure economic sense to do telecommuting, than it has been in the past. The integration of both telephony and the computer technology makes this happen. As the car changed our world 90 years ago, and forced the distribution of people from the core of cities to the suburbs, telecommuting and virtual distributed call center technology will change the distribution pattern of where people live and work and spend in their communities. Many companies will be able to retain their best employees and increase their profitability at the same time. Distributed work environments are now a reality.

Vital Statistics

The "Virtual Distributed Telecommuter" - Georgia Power	
Cost	*Statistics*
Weeks of labor to install system	
Initial System Cost	not reported
Est. Ongoing support Cost	reduced
System Size	
# Ports - Total/# Fax.	
# Users	28
# Calls/year - Teloquent - Atlanta call center	3.5 million
Computer	Distributed Call Center (Pentium system)
Operating System	SCO OpenServer 5.0
Industry	*Power Utility*
Estimated Return on Investment	not reported
Cost Savings	
Manpower equivalency per year saved for staff not commuting to work each month (28 people spending 1.5 hours per day commuting)	4.84 man years
Estimated Manpower savings per year	

The "Virtual Distributed Telecommuter" - Georgia Power

Cost	Statistics
Reduction in costs from legacy PBX "shelf" systems.	20% or about $19,600 for the 28 telecommuters.
Earnings	
Generated new income/year	not applicable
Ripple Effect	
Estimated cost Saving to customer base	ASSUMED 5 seconds increase in throughput, about 170 mandays. 28% of the calls handled by Teloquent users.
Vendor	**Teloquent**
Product	Distributed Call Center
End User	Georgia Power
Computer Telephony Technologies Deployed	
Interactive Voice Response integration	Yes
ISDN based distributed call control	Yes
Screen Pop - legacy systems	Yes

Table 4-2 The "Virtual Distributed Telecommuter" - Georgia Power - Vital Statistics

---

[34] Author's Note: The following statement about this case history by Jerry Cohen of Georgia Power is about the currency of the implementation at Georgia Power of the Teloquent system. Jerry indicated that the Teloquent system was very adequate for the environment that Georgia Power had at the time when the interview for this case history was performed, and is a very powerful system today. It is the author's opinion that the Teloquent solution is a very powerful technology.

Jerry Cohen wrote the following May 9, 1997. "Since this case history was first compiled, Georgia Power has revised its call center infrastructure to be consistent with the other Southern Company centers. The Teloquent and Teknenkron have been replaced by a single ACD in Atlanta with a remote shelf in Macon. The two satellite offices and the work at home agents are supported via remote dial-in units. According to Jerry, while there is a drop in functionality with the new configuration, moving towards a uniform ACD arrangement company wide was a higher priority. It was simpler to replace the Georgia Power setup than to change out the equipment at four other call centers."

Predictive Dialing - maximizing customer person to person contact

P redicitive dialing is another type of technology in a call center or computer telephony workgroup, which greatly increases the amount of personal contact between each person in your team and your customer[35]. There is a basic problem with people dialing and trying to make contact. There is a high probability of unsuccessful attempts before they are able to talk to a person in real time. A predictive dialer is software that is provided with a list of telephone numbers. It dials a number, and if there is no answer, it schedules that phone number to be dialed at another time. If there is an answering machine or fax machine, it can detect what actually answered, and log that in its call record database. If a person says "Hello", it can detect that a person is there, and immediately screen pops a sales script, or customer history to the next available agent in the team, and then connects the person on the phone to the team member. Those calls that are unsuccessful are rescheduled and tried again at a later time. As a result, the team member spends more time talking to real prospects or customers, and spends more time selling, or finding out more about the needs of that customer. This is logged into a sales tracking database, and the database contact can be used again to further customers along to sales closure. It is generally reported that predictive dialing technology can increase the person to person contact time as much as 300%. The overall benefit is that less people are needed to contact a much larger number of customers.

Predictive dialing technology is an important tool in telemarketing and market research and other industries. It also has an unnecessarily bad reputation. It is has been used by telemarketers, who may disturb your evening or weekends with unsolicited offerings of products or services.

One of the challenges in today's markets is having a close relationship with your customers. It takes far more money to obtain a new customer than it does to retain the ones you already have. If your firm sells a commodity, or near commodity item, then the competition can be fierce. The major thing that distinguishes you from your competitors is customer service, and the quality of the marketing of your products. This is how many companies have been able to survive in difficult economic climates. They have retained and even increased their market share, at the expense of their competitors.

Predictive dialers are sometimes used in conjunction with traditional inbound call centers, to allow for the best of both worlds. The software keeps track of the inbound call volumes. Whenever there is a lull in the inbound calls, the predictive dialer software gets into action and starts to dial out. This increases the productivity of agents who would otherwise be either idle, or logout of the inbound call center function and start manually dialing customers or prospects.

Predictive dialing systems provide extensive campaign management and reporting features, that allow managers to monitor call volumes and intercept calls or aid agents, while addressing customer problems. Workflow changes can be made by instructing the predictive dialer software to reroute calls to another agent team, while the original team starts taking a larger volume of inbound calls. The reporting systems provide detailed reporting of time studies and calls completed that are important management tools. The products integrate with existing application software that a company already has, so that information does not have to be duplicated.

In this section we will examine how various industries have been able to increase their sales, and increase customer retention, using predictive dialer technology. These are not your average telemarketing organizations. They take full advantage of the operating system and the networking resources they provide.

EIS International

EIS International of Stamford, Connecticut provides predictive dialing technology on the SCOOpenServer 5.0 platform. The company has been providing such solutions since 1989, and has about 30% of estimated market share[36] .

In the cable television industry, the growth of the market has matured to the point where two-thirds of all American homes have cable television. There is competition from other services and attractions. This makes it even more imperative for this industry to contact their customer base using predictive dialing telemarketing systems, rather than by traditional direct mail marketing methodology.

The rule of business, that it cost far more to obtain a new customer than keeping the ones you have, truly applies to the cable television industry. One methodology to increase customer loyalty and to increase sales, is to provide more added value services to existing accounts. The best way to make that contact is via the phone, using predictive dialing technology.

Comcast, a Baltimore, Maryland cable television provider services two States, said that by using the predictive dialing technology from EIS, they were able to double their live "talk-time" with their customers, from 12 percent to 25 percent of their campaign sample. Even for generating new business, the predictive dialer technology has reduced the cost of sales from $40 to $18 per sale. Another EIS customer in the cable television industry is Adelphia Cable Communications of Couldersport, PA. They are is able to contact their 2 million customers at least 2 times a year with only 22 stations. They did not have to increase their staffing in order to make these contacts.[37] Similar efficiencies are also described in market research[38], and delinquent student loan repayment applications.[39]

Davox Corporation
Davox is another industry leader with large market share of predictive dialing, that utilizes the Sun Microsystems Sparc platform on Solaris 2.4 version of the UNIX operating system.

The Dreyfus Corporation's experience of using Davox's Unison product for predictive dialing demonstrates the efficiencies that can be obtained using this technology. Dreyfus corporation is the 6th largest mutual investment firm in the United States, and manages over $80 billion dollars in assets. It employs more than 450 telephone representatives in its Service Corporation, who sell securities and provide customer service, which receive over 2 million telephone calls per year. Within the Service Corporation of Dreyfus, is the Dreyfus Retail group, which consists of 85 sales consultants who field calls from various leads generated in print and electronic media.

In order to maintain the lowest possible on hold wait time for any inbound lead, call center management at Dreyfus staffed the inbound call group to 98% of capacity. The downside to this was that these sales consultants were not occupied all the time with inbound calls. The inbound calls could fall to 50% of the staffing during lulls in the inbound call rate, the sales consultants would then start to do follow ups on leads and prospect by manually dialing. In order to increase the contact time, keep the consultants working full time, and tap into a rich base of potential business that was inside their existing database, Dreyfus decided they needed to do call blending, using Davox's Unison product.

In this case study, there were a challenges in the merger of an existing sales tracking software called Teletrak with Davox's Unison product. Investment counselors spent only one-third of their time talking to a live prospect or customer. On average manual dialing produced only 3 or 4 contacts an hour.

The benefits derived from the call blending system, and having the Unison system make contact with prospects and customers has increased to 10 to 12 contacts per hour. This is performed by about 30 of the 85 counselors whenever the call volume permits. Dreyfus indicated that an exponential increase in sales has occurred with the 300% increase in customer contact. Customer surveys have shown that 85% of the customers that are contacted using this method welcome the call from their Dreyfus representative. In one special project, managers commented that predictive dialing technology has enabled them to use "one third the people, and complete the project in one-half the time."[40]

Why UNIX for Predictive Dialing?

These industry leaders for predictive dialing technology are on the UNIX platform. The nature of predictive dialing is that it is a multitasking and multi-user process. In effect there are four major processes at work, which are natural extensions to the UNIX operating system's multi-user, multitasking, and scaleable architecture:

1. Database management: UNIX is well known as a platform for its ability to manage large database environments. The database management is twofold in that the campaign list of phone numbers is generated from the database, as well as the need to spawn a database transaction to the screen of the next available agent. This takes advantage of the advanced and mature file systems and networking technology within the UNIX operating system.

2. The next major process is telephony dialing and detection of a real person. This is an intense multi-user function where multiple telephone lines are being dialed constantly, and then spawning sophisticated software algorithms to determine if a real human is at the end the line, rather than an answering machine. This is an example of the utilization of UNIX's multitasking capabilities.

3. Providing multi-user, multitasking services to pop the screen to the next available agent, and then provide the database and screen processing from a small number to a large number of agents.

4. Meanwhile, while all this is going on, another control monitoring process is managing the number of inbound and outbound calls for call blending, so that customer service agents can be evenly paced with customer requests.

While other operating systems can be used for predictive dialing, UNIX's capabilities and reliability have shown it to be the backbone of this segment of the industry.

[35] An excellent source of information about predictive dialing technology is <u>Predictive Dialing Fundamentals,</u> by Aleksander Szlam & Ken Thatcher, First Edition, February 1996 published by Flatiron Publishing 12 West 21 Street, New York, New York, 10010, 1-212-691-8215, or 1-800-LIBRARY, ISBN# 0-936648-80-5.

[36] Article by Stacey L. Bradford as appeared in <u>Financial World, August 12, 1996,</u> 1328 Broadway, New York, NY 10001, Copyright 1996 All rights Reserved.

[37] Case History: Cable Television: "As growth slows, Cable Industry Turns to New Technologies", p.p. 1-2, Copyright 1996, EIS International, Inc. Marketing Department, 1351 Washington Boulevard, Stamford CT 06902.

[38] Case History: Marketing Research Services Inc.: "Predictive Dialing Helps Market Research Company Win New Business, p.p. 1-2, Copyright 1996, EIS International, Inc. Marketing Department, 1351 Washington Boulevard, Stamford CT 06902.

[39] Case History: Sallie Mae: "Sallie Mae Achieves 'Strong' Gains With Two Types of Automatic Dialing", p.p. 1-2, Copyright 1996, EIS International, Inc. Marketing Department, 1351 Washington Boulevard, Stamford CT 06902.

[40] Dreyfus case information comes from 2 sources:

1) Sun Microsystems Inc., 2550 Garcia Avenue, Mountain View, CA 94043-1100, : CMS Success Story: The Dreyfus Corporation, written by Fred Alvarado (415-336-7500). Copyright 1996 Sun Microsystems, All Rights Reserved.

2) Davox Customer Profile: Dreyfus Corporation, Copyright 1996, Davox Corporation, 6 Technology Park Drive, Westford, MA 01886. Tel: 1-508-952-0200, Fax: 508-952-0201.

How does UNIX based CTI work?

In this section we will review the basic building blocks of computer telephony integration, and then delve in how the signaling is achieved between the computer and the telephone networks.

The Three Basic Building Blocks for CTI

The nature of "screen pop" in CTI functionality is essentially the same for a small CTI workgroup of 5 inside sales or customer service agents working on a LAN to very large call center solutions of hundreds of customer service agents as in the Federated Stores FACS case history in this chapter. The same building blocks apply for the most proprietary PBX and computer system integration to the open standards based UNIX PC switch technology.

There are essentially three basic components to allow a CTI application to exist.

1. An up to date and viable database and applications is an absolute necessity. This is my first priority over all others. All the technology in the world does not mean much unless customers can quickly and easily get to their information. Without good database integrity, with efficient ease of access, then all other components of computer telephony integration are not effective. The key issue for database applications such as financial, purchase history, or customer preferences/complaints is, if they are not easily accessible for the customer service representatives, then the more difficult it will be help support your customers.

   In particular, is the problem of "stove piped" applications from different platforms and databases that are presented on a single system. Stove piped applications are different kinds of applications which only share the screen. There is no coordinated database sharing of information. A look up of a customer number in one application, does not preserve the pertinent information, which needs to be keyed in again by the customer service representative in another application. This defeats the reasons for CTI, since the customer will be on the line longer. The solution is to make sure the customer's information can be as quickly and easily access in the least amount of time as possible. This may mean setting up client-server relationships using a open database access standard such as ODBC to allow customer service representatives to have local application client front ends that eliminate the data duplication and data entry times.

Moreover, proper database design and access, can allow management a better decision support for what kinds of customers need what kinds of services. Sometimes, a data warehouse is used to collect large samples of customer information to identify support or buying trends, so that management can be proactive to customers requests. One method of solving the problem of "stove piped" applications is to create a database of databases that links the different customer sources, that reside on disparate corporate platforms (LAN, mainframe, and UNIX systems) so that key information is at hand for a customer support and sales efforts.

While another popular solution is to purchase "near-off-the-shelf" database solutions for applications such as help desk, sales and customer service support. One company that does this is Scopus. A major challenge facing some database administrators is to amalgamate all the disparate database applications using in-house programming resources. It could be either cost prohibitive or take too long to program, test, and deploy these applications from scratch. Scopus's SalesTEAM, and SupportTEAM products provide a completely table driven capability to accommodate the different fields of data from various databases using ODBC (Open Database Connectivity) standards. Depending upon the database on the legacy system, ODBC provide the solution to connectivity to databases on older mainframe hosts. All this corporate data is amalgamated into a single set of table driven screens. The screen sets are platform independent, meaning the screen designs are the same content either they be X-Windows, MS-Windows, and even character based. The use of WebTEAM provides a direction where in a call center, a user can who is viewing a Web page, can request a call by a service agent. The screen is then popped to the agent in the call center, and the call is placed to both the agent, as well as the viewer of the Web page. Moreover, Scopus has incorporated the telephony integration using popular middleware products such as Genesys Labs T-Server. This then allows the database administrator and company management to quickly deploy effective customer service solutions. Scopus is one of many players in the UNIX market.

The key thing about database design, is that the database is used to track when and where the next screen transaction is going to be sent to the customer service representative. This is discussed further below.

2. Voice switching mechanism such as a PBX in concert with maybe an IVR system. This covers the telephone aspect, and allows for customers with an IVR application to help identify themselves if ANI and DNIS information are not provided on the telephone line.

3. A method of communicating between the PBX and the database applications, so that the customer information can be efficiently presented to the customer service representative. There have been various ways of doing this. Switch manufacturers have for years provided special proprietary products that allow for this to happen. This is either achieved by a PBX manufacturer specific interface or recently published standard interface such as TSAPI, TAPI, SunXTL and JavaSoft's JTAPI, and CSTA or via middleware based software companies, such as Genesys Labs, and Dialogic's CT-Connect product on SCO UnixWare. All these models have a single common thread. The ability to synchronize and coordinate information from two different networks. These are the switch voice network, and computer networks.

Screen Pop Management using the Database

No matter how the method the signaling between the two networks is completed, the same database feat is being performed.

Agent's UNIX Terminal device.	Agent Telephone Extension	Automatic Number Identification (ANI) from telephone company	Customer No. (same as phone number)	Customer Priority (1 low to 5 high)
/dev/ttyI1a	557	555-1234	555-1234	1
/dev/ttyp03	347	555-9876	555-9876	3

Figure 4-5 Relational Database intersection table used to control CTI screen pops.

Customer calls, and identifies himself either automatically using the automatic number identification (ANI) and dialed number identification service (DNIS) information, or through an interactive voice response (IVR) application. They want to talk to a human about their problems. The PBX has to determine in its pool of available agents, who is next to listen to the caller's tale of woe, or eagerness to buy. It finds an agent. It signals (there are many ways to do this signaling) to a UNIX system that they have free agent at extension 347, and here is the ANI information about the customer. The UNIX system looks up in its customer table of its relational database using the customer's phone number as part of the SQL statement. In this case, the company identifies the customer by their residential telephone number, which happens to be the same number the customer is calling from. (This may be convenient for this example, but in real life, the ratio of matches from the ANI to the customer telephone number in the database is quite high). The UNIX system starts a database transaction to have its output routed to a screen at a TCP/ip network location /dev/ttyp03, while at the same time signaling the PBX to transfer the caller to extension 347. Ergo, the voice and the data are synchronized and delivered. "Hello Mr. Smith. Are you calling about your delayed shipment to Timbuktu? Don't worry, it was just delivered. Would you like to have the proof of delivery immediately

194

faxed to you?", says the customer service agent, as he reviews the shipping information screen about this African city.

PBX to UNIX System Signaling Methodologies

There various ways this kind of signaling between the voice and data networks. The major differentiation has been how open, and how much has the UNIX computer system been able to encapsulate the voice switching capability.

➢See "The Shore Connection " - WJG Maritel case history, on page 221 as an example of a Call Center PC Switch system.

At one end of the spectrum, PBX manufacturers who until recent history, have offered bridges or software development programs that allowed computer systems to signal to their PBX counterparts through licensed and proprietary interfaces to the PBX architecture. In the middle of the spectrum is the adoption by switch manufacturers as well as middleware developers of market driven standards that allow for easier communication between the two networks. To the other end of the spectrum such as PC Switch systems that totally encapsulate the switched voice component with UNIX standards and applications.

PROPRIETARY PBX SIGNALING

First let us take a look a more classical screen pop integration model from Mitel circa 1993. A SCO developer could purchase Mitel's MiTAI interface software and card that would reside in a PC running SCO UNIX, and then a corresponding card would then be purchased for their Mitel SX-200 PBX. A cable would extend from the card in the PBX to the card within the SCO UNIX system. The Mitel PBX would monitor the telephone extensions of agents in the call center, and as soon as one of them became available, it would signal the event and pass information about the caller, (i.e. ANI or DNIS) which agent was now free, to the SCO system, which would in turn spawn a transaction to the agent's screen, and then signal back to the switch to transfer the call to the agent. Nortel, AT&T and other switch manufacturers have provided solutions like this in the past, and now are providing more open standardized interfaces for their switch technology. Their focus is the PBX, making the UNIX system an peripheral device of the PBX.

MARKET DRIVEN STANDARDS BASED OFFERINGS

For the developer who has been selling UNIX solutions since the early to mid 1980s may even feel this telephony standards battleground is kind of familiar. During the evolution of UNIX there have been many "standards wars" ranging from the architecture of the UNIX kernel to the external network communications methodologies. The solution arose in the 1980's as the market and the industry realized that they would be stronger together with harmonized standards, even though UNIX vendors had overlapping competitive environments. The same is bearing true for the telephony industry, and these standards are now evolving as much as the UNIX based standards that we use today. Millions of people are thankful for UNIX based standards such as internetworking protocols that have allowed the Internet to exist.

The drive towards open market standards started in 1993. They have evolved by the recognition by the PBX manufacturers that they needed to open up their architectures, while at the same time computer manufacturers could deliver the critical market mass needed to bring the benefits of large scale and expensive call centers into the realm of smaller groups or workgroups within the corporate structure. What evolved is many different manufacturers aligning themselves to provide the integration. Through these market driven relationships there has developed two kinds of "open application interface" or OAI. These interfaces are described as being either "Horizontal/Office Automation", while the second is "transaction applications".[41] Call Center applications are of the second category. There are about 3 now possibly 4 major standards that have been developed to meet the needs of the market. Of one that seems to be market dominant is the CSTA based standards. This standard in particular has a lot of support by various PBX vendors.

What follows is a discussion of the different market computer telephony integration application programming interface (API) standards that have evolved. Some are not native on UNIX, but can be used in conjunction to a UNIX database application via networking standards. A quick review of these are necessary, since

196

they all play a part in some form of call center or workgroup CTI. Important thing to remember is that these are standards and not products. They specify how the voice and computer networks talk to each other. It is up to the vendors of computer telephony products to write the software necessary to utilize these access methodologies. Each of the following try to solve the common problem of reducing the labour for software developers to build specific interfaces for every kind of PBX manufacturer.

TAPI

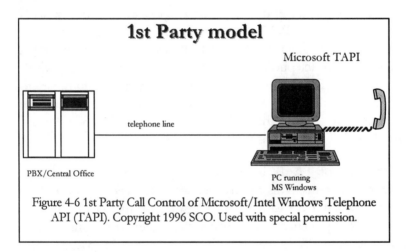

1st Party model

Microsoft TAPI

telephone line

PBX/Central Office

PC running
MS Windows

Figure 4-6 1st Party Call Control of Microsoft/Intel Windows Telephone API (TAPI). Copyright 1996 SCO. Used with special permission.

In 1993 Microsoft and Intel jointly announced with wide industry support of 40 plus software developers of Windows Telephony API. The model shown above in Figure 4-6 is usually associated with TAPI specification. But it can also handle 3rd party call control as well. The TAPI model is known for its "First-party call control" capabilities. "TAPI supports first-party call control only (meaning that the person involved with the call actually routes and controls the call).... First-party call control is typically of greatest importance to individual and workgroup users of telephony services..."[42] This means that the PC running a Microsoft application provides the CTI link but it is up to the individual user to maintain control of the call, such as call transfers.

Why is this significant? It is significant since it provided the mass market support for telephony applications within the PC. This allows the PC to control, to take and place calls, conference, and many other functions from the PC screen, rather than from the complexity of the digital phone. It allows through its API the ability for Windows clients to integrate voice, fax, and other data into a multimedia format. Applications such as integrated messaging, using Microsoft Exchange and others provide the ability to perform tasks that were only available in large telephony environments such as high density call centers with legacy PBX and data systems. It also helps the Windows developer to be isolated from the complexities of the public telephone network standards (i.e. ISDN, T1, and POTS). The operating system provides its own middle interface that now allows telephony developers to develop a broad range of telephony services to the API specifications, which in turn allow for the wider Windows development community to see TAPI as a regular MS-Windows resource. Since 1993 these services have been further enhanced in both the Windows95 and Windows NT operating systems. [43]

In the case of TAPI, the interface is provided in the Microsoft operating systems. This means that software developers on those platforms only have to write the software for the TAPI interface. This is usually likened to the change that occurred for print drivers for Microsoft Windows 3.1. Before in the Microsoft DOS operating system, if you wanted to use a printer to print a document, you had to select a print driver supplied by the application software developer (i.e. Lotus 1-2-3). This was extremely system administrative intensive, and it forced every software developer to have to maintain their own library of print drivers and ship it with their products. This was solved in Windows 3.1 where the operating system maintained the print device drivers and supply software developers with a specification to send data to the printer interface Microsoft supplies. The same holds true for UNIX operating systems as well. System administrators of UNIX systems can provide a single interface to all print requests from native UNIX applications and as well through print request from networked PC's or other computer systems.

Why should TAPI be a concern for a UNIX based solution provider? Beyond knowing how and what the technology is there is a real reasons for being knowledgeable about this standard:

1. In a TAPI model it is possible for the UNIX system and its database capability to be a major data resource for a TAPI enabled Windows 95 or Windows NT Workstation system. In this architecture, the call arrives in the PBX, the call handling is directly signaled to the client or through a Windows NT server to the client, and the client can then haul up a screen transaction of customer information from the UNIX database using OBDC based technology and networking technology.

TSAPI

TSAPI (Telephony Services Application Programming Interface) is based upon the CSTA API (see below). It was developed by Novell in cooperation with AT&T. TSAPI provides a standardized method using a Novell server with a special Network Load Module (NLM) to communicate with the PBX and the Novell server and the Novell PC clients. This installation requires Novell's Netware Directory Services[44]. One major difference between TSAPI and TAPI, is "The biggest reported advantage of the Novell approach is that it reduces the developer's learning curve while allowing the legacy systems of the telephone industry to participate in the CTI revolution.....while TSAPI adds third-party control, wherein an agent" (i.e a computer term not a human) "or application that doesn't 'own' the call can control the routing".[45] The initial advantages of TSAPI when it was announced in 1993 was that it is server rather than client based such as TAPI initially was designed. This provided an advantage for both screen pop and as well as client software management of call handling. TSAPI has had its own set of PBX vendors that have come out to support this API. Even some PBX vendors have tried to develop their switching software to meet the needs of both standards.

How does this concern a UNIX developer and/or purchaser of a computer telephony application? Again, the issue is that the UNIX system can play a roll as a database server to a Novell server for "screen pops" A call arrives to the switch, and then the next available customer service representative can be identified, and via networking protocols such TCP/ip or even IPX/SPX a database transaction could be spawned on the UNIX server to appear on a Novell client, or possibly another TCP/ip based client. Again, like TAPI, the UNIX system is not necessarily providing the traditional call control functionality, and has to rely on another client of it to perform this necessary task.

CSTA

Computer Supported Telephony Applications (CSTA) is supported by a growing number of PBX vendors such as Siemens Business Communications Systems, Alcaltel and Ericsson.[46] It is referred to in the literature as a CTI link as a protocol rather than an API. CSTA is a standardized protocol put forward by a industry body called European Computer Manufacturers Association (ECMA). Unlike the TSAPI and TAPI CTI standards this standard is proposed by a standards body, and not a manufacturer with vested market interest. Like the other standards it provides for a methodology of signaling between the two networks.

SunXTL Teleservices

SunXTL Teleservices is from Sun Microsystems. This is a software technology and API proposed by Sun to allow for the development of distributed enterprise client-server computer telephony solutions. It consists of a software toolkit that allows a Sun Solaris developer to obtain a buffer between the application layer, and the telephony services layer.

The following are the goals of SunXTL Teleservices

1. "Provide a development platform for call center management applications.... Need to integrate the ability to control PBX call control functions into a robust, distributed computing environment.

2. Provide a development platform for desktop CTI applications. Applications such as 'Feature Phone' GUI, remote access via

DTMF, personal voice mail etc. can share access to a workstations telephony hardware

3. Provide transparent porting between PBX, analog, ISDN, ATM and other communications media technologies.

4. Provide simple access to common voice services.

5. Enable specialized or non-voice services."[47]

SunXTL solves this by providing software and architectures to meet these goals. It protects the UNIX software developer from some of the hardware specialization that are presented when dealing with different kinds of telephony medium such as analog DTMF, digital technologies such as ISDN and ATM from the PBX integration.

You can envision the SunXTL architecture as a three layered platform. At the top layer is the computer telephony application that a software developer using Sun Solaris 2.4 or later has used the SunXTL libraries in the development process to gain access to the telephony services. The middle layer consists of the SunXTL server which maintains its own database of information about the lower layer SunXTL providers. The server is the centralized keystone component from which security, and access to the Teleservices, and resource management. The lowest layer is the Media Programming Interface which contains the providers. The providers in turn interface with the telephony hardware such as ISDN, analog, and ATM platforms. The top layer has different XTL library than the lower media programming interface requires.

Here are the 4 major components of the SunXTL architecture:

1. "SunXTL Libraries. These are the object-oriented programming interfaces for the programming interface layer, and the media platform interface.

2. SunXTL Server: multi-client & multi-device support for Teleservices, resource management and security.

3. One or more Providers: Providers are the device-specific software entities that enable telecommunications device to communicate with and/or be controlled by the SunXTL platform and its applications.

4. A Data Stream Mulitplexor. The universal mulitplexor (Umux) provides a uniform means for applications to access and share data channels associated with a telephone call."[48]

Summation of Market Driven CTI Application Programming Interface Standards

There has been two forces at work that have allowed and enabled the market driven standards for CTI.

1. The move from large mainframe centralized applications to client-server architectures. The software applications have to be engineered to meet the needs of computer-telephony. In some large mainframe applications were not suitable for computer telephony integration, even though the return on investment was very attractive. The advent of client-server applications and 4GL distributed database systems have allowed for the easier deployment of computer telephony applications.[49] This can be seen in the use of the Teloquent installation for remote clients for Georgia Power in the case history in this chapter. They found that it was more efficient to support remote in-house or remote offices because of the ability to use ISDN and individual client-server systems, than it would have been to install all the communications hardware for a CICS database application and as well the telephone circuitry. They could do it with the older system, but the incremental cost would be much higher.

2. Economics: The CTI automation has a proven track record of providing an identifiable economic benefit. Hence the need to provide this to a large standardized computer market has been the major impetus to the development of market driven standards.

All of these standards are evolving and each address different parts of the computer market. It is interesting to see how the PBX vendors are trying to placate the various computer market champions such as Microsoft, Novell, Sun and SCO. As an owner of a PBX or planning to purchase a PBX then you may feel compelled to pick one PBX because of its dedicated or shared support for one or more market driven telephony APIs. This may force you to purchase a system other than UNIX to support these features. But all is not lost for the UNIX developer and UNIX customer. The common philosophy of the UNIX market is standards and choice. There are de facto standards which are rules which have been laid down by a major vendor, and the market has accepted *en masse*. A good example of such a standard is the Microsoft Windows interface, or DDL, or OLE technologies. Then there are standards that are designed by multiple vendors with a common vested interest to help broaden the market and these standards are developed through end user input and vendor consultations. Examples of this type is the X-windows standards and as well as Spec 1170 for the 1170 UNIX system calls that are now part of the "unified" UNIX kernel. The UNIX community goes further by placing the responsibility and certification of specifications into vendor independent organizations such as X/Open.

As for the telephony API standards, they tend to be provided by vendors of vested interest, and for which the market has not necessarily chosen a predominate or champion standard over any other. Hence the development of "middleware" technology that provides a single interface between the software developer of UNIX telephony applications and the various kinds of market driven telephony standards and proprietary PBX interfaces.

MIDDLEWARE TECHNOLOGIES

A middle layer of software integration exists from companies like

Customer service/purchasing Application
Middleware Bi-directional gateway between the database and voice networks using and middleware and/or market driven standards such as TSAPI, TAPI, CSTA or Proprietary PBX interfaces.
PBX interfaces

Figure 4-7 Dipiction of the role of "CTI Middleware"

Dialogic and their CT-Connect product, or Genesys Labs T-Server. These are the middleware products, that provide UNIX CTI developers software solutions that provide a buffer to all the changing standards, and proprietary PBX interfaces. These software packages are valuable in a changing and evolving standards market such as computer telephony. UNIX database application developers want to CTI enable their applications with the least amount of worry of how and why a PBX works. Since their UNIX world has many hard fought standards, they do not to want to peg their hopes and dreams to necessarily to a single PBX vendor or one market driven standard that seems dominant at the time. They want to sell their solution to the widest possible market, and they do not really want to be bogged down to the nuances of every PBX. They want to be told when the application needs to send a transaction screen to a particular terminal, and what information it should contain, while simultaneously signaling the PBX to connect the caller. In the case of predictive dialers the computer system tells the PBX to contact an individual at a specific number, and when there is a real person at the other end, then the

UNIX system tells the switch to transfer the caller to the next available agent, and then it pops the screen to the caller.

There are different vendors on UNIX that provide this middleware technology. I have highlighted three below, but by no means the list is exhaustive.

CT-Connect - Dialogic's product on SCO UnixWare

On October 8, 1996 Dialogic and SCO announced that jointly the companies would port Dialogic CT-Connect middleware to SCO UnixWare. CT-Connect is a product that has been used by developers for some time on other operating system platforms in order to manage the integration between the voice and computer networks. By the end of the second quarter 1997, CT-Connect will provide support for TAPI and DDE with subsequent support for TSAPI. The product will also support the proprietary interfaces form Lucent Technologies DEFINITY PBX(CallVisor/ASAI), and Nortel Meridian 1(Meridian Link). Furthermore, CT-Connect for SCO UnixWare will support the "CSTA link protocol used by Alcatel, BBS Telecom, Bosch Telenorma, Cortelco, Ericcson, Intecom, Mitel, Rockwell, Siemens Business Communications Systems, Tadiran and other switches."[50]

The main immediate benefits for this product release are that for CTI developers now can access directly the major telephony APIs directly from SCO UnixWare, which will greatly simplify the solution set complexity, and reduce the cost to the customers of the solution. By supporting TAPI and DDE (Dynamic Data Exchange) capability this obviates the need to network a UNIX system to an Windows NT server for PBX control.

The product will be offered in four configurations:

1. "A Full configuration, supporting all types of client modules and an unlimited number of client systems;

2. A Desktop configuration, supporting an unlimited number of client systems but restricted to desktop applications

3. Desktop Lite configuration, supporting up to 36 client systems running desktop applications

4. An Evaluation configuration, licensed free of charge, supporting up to 16 client systems running desktop applications. "[51]

Some benefits of CT-Connect on SCO UnixWare will mean to the UNIX community, especially to the large UNIX community running on Intel's processors are:

1. New markets: Access for UNIX developers to new market segments that they may not have been able to exploit natively on UNIX. A good example of this is TAPI support and other APIs.

2. Ease of development: Developing for one interface rather than many interfaces will simplify and lower the costs of development.

3. Choice: Of interest is the ability to write your CTI applications so that you can move amongst the different CTI APIs. This means as the market shakes out the "CTI API War", the developer wins in the end, since the investment protection with a middleware product will allow for the movement of the application to different APIs in the market. One of the major tenets of open systems standards is choice. CT-Connect and other middleware telephony solutions provide the customer choice.

T-Server - Genesys Labs Product on Sun Solaris

Genesys Labs of San Francisco, CA has developed T-Server as their core CTI middleware software. It runs on Sun Solaris platforms on Sparc systems.

T-Server is a tried and true middleware performer and has been installed at customer sites where there are multiple call centers supporting over 1600 agents. The software is able to route and manage both the communication between the server and databases on the LAN or WAN, and as well maintain the flow through of call identification by ANI or IVR input. The calls can be "screen popped" to many different agents, and the screen transfer between agents. The product can handle more than just a voice traffic and the coordinated delivery of data screens. It can also handle video-kiosk applications, and many other kinds of media. T-Server can be also used for call blending for outbound call support.

"Once again our T-Server spots that it's a video call and intelligently routes that medium along with voice down to the appropriate agent desktop, where they can also do voice, data and video transfers...... it's a greatest way to bring in a selling specialist in retail outlets. When shoppers need special help, they walk over to one of these things, pick up the handsets and they're connected live to a visual agent who knows exactly how to deal with them."[52] This statement was made by Gorge Geros, VP Business Development of Genesys Labs.

T-Server communicates with different kinds of APIs and or proprietary switch interfaces from

- Aspect
- AT&T G3
- Nortel M1
- Nortel DMS
- Rockwell Galaxy
- Rockwell Spectrum
- IVR Systems from suppliers such as Nortel, AT&T, Syntellect, Brite, Voicetek, Edify, InterVoice, Periphonics.)
- New Interfaces as they are custom built.

The communication with the client technology is via Cobra, OLE 2.0, Genesys Labs' own T-Library, TAPI, TSAPI, and other new standards as they are released.[53]

T-Server is noted for its robustness and open architecture. This is how the aforementioned vendors and APIs have been able to communicate with this platform. T-Server is the "traffic cop" that helps to be a cornerstone for 3rd party application software such as

Scopus or developed by in-house development teams or for other products, as well as supporting the host of other Genesys Labs software. These include Call Center Manager which monitors real-time events, Campaign Manager that manages both dialing and transfer instructions, as well as voice processing results and real-time telephony events. Call Router that manages the call handling using information obtained via ANI, DNIS, and IVR information, and provides call routing instructions. Call Concentrator is used to manage real-time telephony events between linked call centers located in different parts of the country or possibly planet. Stat Server for statistics of the real-time telephony events.

JTAPI (Java Telephony API) - JavaSoft

With the heightened interest and demand for world wide web applications, the Java programming language allows for applications to be distributed through the Internet, and are loaded on demand from a remote server anywhere in the world. The world wide web is well known for publishing and linking of documents throughout the world. In order to process orders and provide transaction processing, the Java programming language provides the means to load applications and execute them on the local processor. Moreover, the processor in this case can be Network Computer (NC) or any other computing device (including PCs) to run applications. The difference in cost of an NC versus the cost of a PC is projected to be around $500 to $750 for the NC machine, while the PC is averaging retail price is about 2,000 to 2,500. From a corporate network management point of view the promise of reduced costs system administration costs of the NC machine versus a PC is dramatically different. Since all system management and application management is located and loaded from a server then there is a labor reduction in the cost of loading and managing software. Since software is uploaded from a Internet server, then there is the best case scenario for software release management. NC computers do not even have mass storage hard drives and floppy disks, which even increases the ability of the software management. There is a reduced threat of end user unintentionally induced computer viruses, using this model. Proponents to the NC model have used a recent Gartner

Group report about the cost of servicing and maintaining PCs in a network environment. The report stated that 12,000 per year is the average cost per year per PC.[54] This includes the capital costs, and manpower cost.

Java as a language is platform independent and this includes computer systems and even other devices such as phone systems and other electronic systems.

JavaSoft, a Sun Microsystems Inc, business, proposed JTAPI as a Java based telephony API. This is an open public specification that was available for review until December 30, 1996 for input. The stated goals of this telephony API are as follows:

"The Java Telephony API specifies the standard telephony application programming interface for computer-telephone applications under Java. It provides the definition for a reusable set of telephone call control objects which enable application portability across platforms and across implementations.

The Java Telephony API represents the combined efforts of design teams from Sun, Lucent, Nortel, Novell and Intel, all operating under the direction of JavaSoft. It insures that a simple interface is available that can be reused and extended for a broad range of computer-telephony tasks."[55]

Because of the ability of Java to run on any computing device, and a Java application can be retrieved from any other computer on the Internet with permission, means that Internet based telephony applications could be viewed as a common interface to the other API's already mentioned in this section. In fact the stated goal of the JTAPI is to have "Applications writing to the JTAPI are independent of platform or phone system. JTAPI interfaces will soon be available for other computer-telephony integration applications, such as SunXTL, Microsoft and Intel TAPI, Novell and Lucent TSAPI and IBM Call Path".

The Internet seems to be a great leveler of the computing industry of late. It is composed of many standards which have had their origin in the UNIX marketplace. The Internet has been described as democratized anarchy, where no one entity (i.e. person, place or thing) maintains absolute control. This means that Java based telephony integration is a great harmonizing effect.

The proposed JTAPI specification proposes two kinds of configurations:

1. Network Computer (NC) Configuration

In this configuration JTAPI runs on a remote workstation communicating to a telephony server. The remote server can be communicating with the PBX or digital connectivity to the public telephone network using ISDN or even ATM, or interestingly to an Internet protocol. The end user of NC workstation is able to manage calls using Java application software, and is insulated from how the telephony API is working or is designed. Thus the same Java application can work with many different kinds of telephony API's whether they be from Sun, Novell, Microsoft & Intel, or proprietary interfaces such as Nortel's Meridian Link. Moreover, once the Java application is tested it can be run on many different platforms. It may be possible in the not too distant future to see a

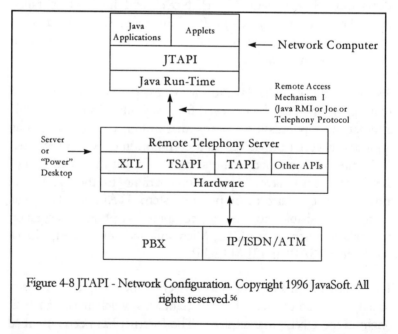

Figure 4-8 JTAPI - Network Configuration. Copyright 1996 JavaSoft. All rights reserved.[56]

NC system using JTAPI to manipulate call handling for a call center or workgroup as deftly as would a Novell server could for TSAPI, or TAPI based Windows NT system.

2. Desktop Computer Configuration

The desktop configuration contains all the telephony services and

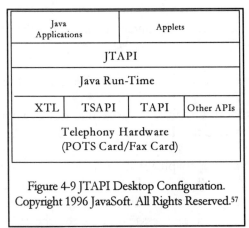

Java Applications	Applets		
JTAPI			
Java Run-Time			
XTL	TSAPI	TAPI	Other APIs
Telephony Hardware (POTS Card/Fax Card)			

Figure 4-9 JTAPI Desktop Configuration.
Copyright 1996 JavaSoft. All Rights Reserved.[57]

hardware within the same chassis.

Within the JTAPI specification there is a "core package" with 8 "extension packages" and these can be increased to help with the extensibility of the API. The core package looks after basic telephony applications such as adding telephony capability to a Web page. The extension packages are:

1. Call Control - such as call transfer, conference, pickup and park, as well as reporting functions that elaborate on a call's connection status.

2. Call Center - provides for ACD, call routing and other functions associated with large call centers.

3. Private Data - "Allows applications to communicate provider-specific data to the telephony subsystem"

4. Terminal Set Management: "Allows applications to control the physical features of a telephone set, such as speaker volume, lights and LCD."

5. Capabilities - "Allows applications to query the JTAPI on its ability to perform certain tasks, for example whether it can transfer a call or set up a call forwarding feature for a particular phone number".

6. Media Services - "Handles audio and other media from a telephone line"

7. Mobile Phones - "Provides support for cellular telephones."

8. Synchronous:"Enables the system to run in synchronous mode as opposed to the default asynchronous mode".

Of interest of this chapter is the discussion pertaining to the Call Center Package. Java as a language is object oriented. The Call Center Package consists of different objects that in turn manage and look after such as messaging between the PBX and the Java application, the access to the ACD functions of the switch, Agent Terminal States (i.e. ready, login, logout, busy, and not ready, and work ready), and Predictive call states.

JTAPI as an API is strong as the APIs that have been developed by the other major API developers (CSTA, Novell, Microsoft and proprietary links). This means that new APIs can be used and developed to make the use of JTAPI extensible and evolutionary. However, because of the use of Java based technology already has a huge installed base of millions of computers by the fact that most of the popular browser technology from Netscape and Microsoft can run Java programs, this means that the ability to run telephony applications such as complex call centers to even integrated messaging will be made much easier. Moreover, software developers of call center applications can write to the Java language and JTAPI interface and then instantly port the application to many different computer systems. The source code of the application can be easily managed from one or many servers on the Intranet or Internet.

PC Switch - The Integration of Voice into UNIX Computer Networks

From a non-traditional PBX point of view the PC switch provides the merger of both worlds of standards based technologies within a computer and telephony inside a single chassis. There is no need to provide special signaling between the telephone network PBX to the computer network, other than the UNIX operating systems standard signaling methods such as inter-process control (IPC). This provides an very interesting, flexible, scaleable, UNIX based solution. This is where UNIX shines as a platform, since it has all the inherent multiple protocol computer based networking, and database support, along with its ability to support the telephony services. Its multitasking and multi-user functionality and structures are well suited for this task. After all, UNIX was developed originally at AT&T Bell Labs for telephony and scientific applications. It only took the computer revolution to force the convergence and the standardization of the products inside the UNIX chassis to allow this reincarnation to surpass its evolutionary branch of the family tree - the proprietary PBX. This is fully described in Chapter 5 - PC Switch on page 217. The "Shore Connection" case history demonstrates how well this technology can have a significant impact on ROI for the customer, and provide better service for their customer's customer.

For call center solutions, the PC switch is well suited for the CTI workgroup of a handful of staff to a very large system of hundreds of agents on a single system. The uniqueness of some PC switch architectures in call center applications is that they can be easily changed to meet the needs of the call center environment. Unlike other traditional vendors of PBX and call center solutions, some of these vendors are able to customize the software to the exact specifications of the application needs of the workgroup or large call center. This gives them an competitive advantage over the traditional vendors, who have to either provide expensive customizations with longer turn around times, or the customer may have to wait until the next software release for the switch to accommodate the needs they have expressed. The PC switch can communicate with most kinds of networked displays and dial-up terminals on the market. They can communicate in digital formats for both the telephony and data capability. Moreover, they are less expensive since they use standardized hardware and software technology that is in the mass market. So they are easier to maintain.

---

[41] Newton's Telecom Dictionary, 11th Edition, "OAI" entry Page 425, by Harry Newton, Flatiron Publishing, 12 West 21 Street, New York, NY 10010. ISBN 0-936648-87-2.

[42] Office of Information Technology - Boston College Strategic White Paper: Voice Services - DRAFT: Computer-Telephony Integration: http://www.bc.edu/bc_org/tvp/voice/whitepaper.html - Page 4 as of the date of 2/7/97 1:14 PM. Copyright 1996 - The Trustees of Boston College, and Copyright 1995 by CMP Publications. March 1, 1995 Issue 1103, Page 129 Section: Computer Telephony

[43] Newton's Telecom Dictionary, 11th Edition, "Windows Telephony"" entry p.p. 669-670, by Harry Newton, Flatiron Publishing, 12 West 21 Street, New York, NY 10010. ISBN 0-936648-87-2.

[44] SCO Products: Choosing a Computer Telephony Server page 3. Http://www3.sco.com/Products/telecom/white.html as posted on the World Wide Web on the Internet as of 1/21/97 10:37 PM.

[45] Office of Information Technology - Boston College "Strategic White Paper: Voice Services - DRAFT: Computer-Telephony Integration": http://www.bc.edu/bc_org/tvp/voice/whitepaper.html - Page 4 as of the date of 2/7/97 1:14 PM. Copyright 1996 - The Trustees of Boston College, and Copyright 1995 by CMP Publications. March 1, 1995 Issue 1103, Page 129 Section: Computer Telephony

[46] The list of vendors appears in 2 different references: "Choosing a Computer Telephony Server", p.3 http://www3.sco.com/Products/telecom/white.html as of 1/21/97 10:37. 2nd reference: "An Introduction To Computer Telephony", p. 7 http://www.dialogic.com/company/whitepap/carlieee.html as of 2/9/97 3:28 PM by Carl R. Strathmeyer, Dialogic Corporation, Copyright 1997 Dialogic Corporation. Appeared in the IEEE Communications Magazine May 1996

[47] Computer Telephony on the SUN Platform, pp. 108-109 by Patrick Kane, published by Flatiron Publishing, 12 West 21 Street New York, NY 10010 ISBN 0-936648-85-6.

[48] Computer Telephony on the SUN Platform, pp. 109 by Patrick Kane, published by Flatiron Publishing, 12 West 21 Street New York, NY 10010 ISBN 0-936648-85-6.

[49] "An Introduction to Computer Telephony" p.p. 9-10 http://www.dialogic.com/company/whitepap/carlieee.html as of 2/9/97 3:28

PM by Carl R. Strathmeyer, Dialogic Corporation, Copyright 1997 Dialogic Corporation. Appeared in the IEEE Communications Magazine May 1996

50 "SCO and Dialogic Announce the First Open CTI Server for UNIX Systems", Page 2. Monika Laud, as found on the world wide web http://www3.sco.com/Products/telecom/p100896a.html. 1/21/97 10:38. Copyright 1996 Santa Cruz Operations, 425 Encinal Street, Santa Cruz, California.

51 "SCO and Dialogic Announce the First Open CTI Server for UNIX Systems", Page 2. Monika Laud, as found on the world wide web http://www3.sco.com/Products/telecom/p100896a.html. 1/21/97 10:38. Copyright 1996 Santa Cruz Operations, 425 Encinal Street, Santa Cruz, California.

52 Article: "Star of the Industry: George Geros", September 1995 edition of Computer Telephony Magazine. Flatiron Publishing, 12 West 21 Street, New York, NY 10010.

53 Article: "Star of the Industry: George Geros", September 1995 edition of Computer Telephony Magazine. Flatiron Publishing, 12 West 21 Street, New York, NY 10010.

54 "Thin clients appear slim on true substance", Network World: Networking Strategies for the Enterprise, Volume 7, Number 3, February 14, 1997, page 18. Laurentian Technomedia Inc, a unit of the Laurentian Publishing Group. Network World Canada 501 Oakdale Road, North York, Ontario M3N 1W7. http://www.lti.on.ca

55 "Java Telephony API", Page 1, Copyright 1996 JavaSoft, A Sun Microsystems, Inc. Business, as posted on the world wide web url http://java.sun.com/products/javatel/index.html as of 2/7/97 3:32 PM.

56 "Java Telephony API", Figure 1: Network Configuration Page 2, Copyright 1996 JavaSoft, A Sun Microsystems, Inc. Business, as posted on the world wide web url http://java.sun.com/products/javatel/index.html as of 2/7/97 3:32 PM.

57 "Java Telephony API", Figure 2: Desktop Configuration Page 3, Copyright 1996 JavaSoft, A Sun Microsystems, Inc. Business, as posted on the world wide web url http://java.sun.com/products/javatel/index.html as of 2/7/97 3:32 PM.

Chapter 5

PC Switch

A Lower Cost Standards based Voice Switching does exist!

A PC Switch is different from a regular private branch exchange system in that it employs commonly used standardized parts and software technologies built upon an "industrialized" personal computer. These systems have a lot of redundancy built in such as two power supplies, 2 cooling fans, lots of high speed disk drives that use redundant array of inexpensive disks (RAID) technology. This technology allows you to mirror the data from one set of disks to another, allow for faster simultaneous accesses to the data, and have the ability to replace faulty a drive, while the system is still running, and have the data that was lost on the bad drive reconstructed on the replacement drive, from data redundantly stored on the other drives in the system. This dramatically reduces the cost of ownership, and maintenance of the switching technology, and greatly increases the capability, reliability, and flexibility of the system.

Interestingly, the UNIX operating system had its beginnings in the telecommunications industry, when it was developed by AT&T Bell Labs many years ago. Ironically, the UNIX operating system of today provides such a dramatic cost saving, and flexibility as embodied in a PC switch, that this technology is competing very well against the traditional private branch exchange technology. It is because of these factors and successes, the PC Switch is predicted to grow incredibly over the next few years.

◄ See forecast growth of UNIX based PC Switch technology in **Figure 1-2** on page 6.

Besides the growth in market, this represents an fundamental shift of responsibilities and economies within the convergence of computer telephony. The case histories in this chapter demonstrate how fundamental this change is and how voice traffic will be managed. Of special note, is the case history pertaining to world wide Intranet telephony that has been achieved by AlphaNet Telecom. This case history, and the others in this chapter mark the

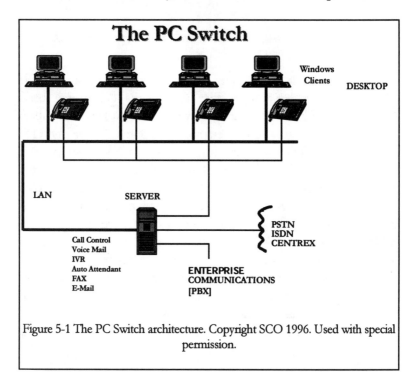

The PC Switch

Windows Clients

DESKTOP

LAN

SERVER

Call Control
Voice Mail
IVR
Auto Attendant
FAX
E-Mail

ENTERPRISE
COMMUNICATIONS
[PBX]

PSTN
ISDN
CENTREX

Figure 5-1 The PC Switch architecture. Copyright SCO 1996. Used with special permission.

point in history and future, where traditional voice switching and management performed by proprietary private branch exchanges is usurped by the computer industry's model. Voice becomes just another type of data floating inside the computer, and is supported and managed with the same principles.

PC Switch technology could be said to be the telephone switch come full circle. What is the difference between a PBX manufactured by a traditional PBX manufacturer versus a PC Switch? There are immediately 3 major differences that come to mind:

1. That the component products are based upon "off-the-shelf" open standards based products. The telephony hardware is based on MVIP or SCSA (or both) technologies. This means that products from Amtelco, Dialogic, Linkon, and others can be purchased to provide both the telephone network interfaces (i.e. Analog, ISDN, T1, and E1) as well as the specialized technologies of text-to-speech, and speaker independent recognition. They also use "industrialized" Personal Computers. These computers have redundancy built in them, in terms of disks, power supplies and other major components that tend to fail in regular PC's used in the home or office. Even these components have been beefed up. The usual refrain from traditional PBX manufacturers is "how can you depend on a PC for mission critical communications? The mean time before failure for a telephone switch is measured in decades." The refrain from PC Switch advocates, is that most major mission critical database applications in corporate, and departmental servers have been successfully run for years. Many new technologies such as clustering of computer systems, automated rerouting or cut over of traffic. Yes, there is a measure of fallibility in even the most successful and resilient systems. The ability of Industrialized PC systems would meet most of the demanding mission critical requirements. If you are really worried about a shutdown of the phone system - they are relatively inexpensive - buy two, and you effectively double the MTBF rating. They also have access to very widely known and low cost networking hardware, and protocols such as TCP/ip in order to provide the completed picture.

2. This also means that software developers who use these standards can build very sophisticated ACD's, conferencing, and other traditional features of a traditional switch. More importantly these software features are based upon open standards based UNIX operating systems and other operating systems. They utilize the most advanced software technologies in the market such as object oriented programming languages such as C++ and Java. By using standard based programming technologies then the cost of development, test, and production products is much easier. If there is a fix that is

needed to be distributed, it is easily done via dial-up process or even downloaded from an FTP site or Web site from the Internet. Changing a single component or feature now is a simple task, rather than making a single change and have to do many man months of regression testing as to happen with older proprietary switching technologies. This also means better and easier remote monitoring and fixing of these systems. Standards such as Simple Network Management Protocol allows system administrators to remotely and quickly diagnose and fix problems.

3. Lastly, the cost factor is much lower than some larger and equivalent proprietary PBX systems. The adherence to standards and the ability to purchase mass produced off-the-shelf products results in a scale of economy that is very hard to match. A case in point is the return in investment in the "Shore Connection" PC switch case history in this chapter reportedly saved as much as $1,000,000 as compared to the same equivalent in traditional PBX technology.

The "Shore Connection" - WJG Maritel

Have you ever wondered how a ship to shore telephone call is made by a Very High Frequency (VHF) radio? This case history reviews how Maritel of the United States, has been able to provide telephone coverage to most ship to shore telephone traffic for inland waterways, Pacific, Gulf and Atlantic coasts. This coverage extends hundreds of miles

Figure 5-2 Cygnus Corporation provides a complete PC Switch & PC based Call Center Solution for WJG Maritel's ship to shore telephone services. Scene above is a dramatization. Copyright 1997 Scheduled Solutions Inc. All Rights Reserved.

off shore. This case history demonstrates how a UNIX based PC switching system developed by Cygnus Corporation is a full private branch exchange, including complete local area networked call center, was able to provide much more than an equivalent solution offered by a major legacy private branch exchange manufacturer, for greatly reduced cost, with increased flexibility and growth.

Background

Picture yourself as a president of a large oil company. After months of traveling around the world, talking to investors, and exploratory engineers, you finally got away with your family for a 2 week cruise exploring the northwest coast of the United States, and then on to Hawaii. No meetings, no decisions, no newspapers and no television. When you left, you told your office not to disturb you, unless it was a dire emergency - if they could find you! You even left your cellular phone and laptop computer where they belong - back at the office. These weeks are for the people you care about the most. On your third day, the ship stops at its first port of call to allow the passengers to take a bus tour through the Coast Mountains of Oregon. Your family enjoyed the outing. After the returning to the ship, it soon departed for a long voyage west to Hawaii. You and your family retire to your cabins to prepare for dinner. As you get ready, you think what a wonderful time you are having. All you can think of is days of fun at sea, and full of expectation of the wonderful things to come in Hawaii. You meet with some new found friends for dinner.

After dinner, the children leave to go to the recreational center. Your wife, friends and yourself retire for an after dinner drink at the bar beside the pool. Your mind drifts off from the quiet conversation to take a good look at the scene that is before you. It's a wonderful warm evening with a large full moon low in the horizon framed by a myriad of brilliant stars, illuminating a flickering dance of light on the waves, creating a shimmering sea. You think to yourself how beautiful it is to stop and actually see what is around you for a change. You are finally starting to relax and unwind.

Your attention now drifts back to the group, and you start to listen to the trailing end of the conversation being uttered by your friend. He said he had a chance to purchase a newspaper while on shore that day and had read the business section. The feature article was about a large oil company battling a hostile takeover bid, and the company's stock had dropped 25% in price, since the president of the company had not made a public statement for 2 days. Suddenly, as you hear her gasp, your wife's hand grabs your arm in a vice grip, making you spill some of your drink. Incredulously, you suddenly realize he is talking about your company! You are 160km (100 miles) west of Oregon! You rush off to the purser's office for help. There goes the vacation! You have to get to a phone! The purser places a ship to shore phone call using the ship's very high frequency (VHF) radio, using the services of WJG Maritel.

Glen Smith, president of Cygnus, describes how WJG Maritel has been in the ship to shore telephone business for quite a while, with a consolidation of 147 very high frequency radio sites: that cover the Pacific, Gulf Coast, Atlantic, and inland waterways of the major rivers in the United states. Soon, Maritel will be providing coverage for the Great Lakes region of the United States.

Maritel manages the communication between the ship and establishing the call with the land based caller, and then conferencing them together. The call is timed and the billing is made to the ship who originated the call. The current systems were not meeting the growth of their business. The Maritel operators were burdened with administrative tasks that sometimes allowed for calls not to be correctly billed. They needed better switching equipment that could handle the increased call volume, and they needed better workstation software to shorten the time it took to setup the land based portion of the call. This would alleviate the administration tasks and increase billing accuracy.

System Overview

Cygnus uses voice processing cards from companies such as Pika for voice processing, Xircom for primary rate interface cards, and Amtelco, for voice processing cards that are Multivendor Integration Protocol (MVIP) bus based. These components reside inside an Intel Pentium PowerStation system running the QNX operating system. For some of these cards, Cygnus wrote the system device drivers to enable them to run on the QNX operating system. QNX is a UNIX operating system that has been modified to provide a real time capability. Real time capability in

◁ For an explanation on MVIP and other voice processing card buses please see page 45.

an operating system is the ability of software to control events or processes, in a manner so that other processes concurrently happening do not interfere with the attention of the central processor (i.e. Intel Pentium) and the timing of the current process being executed. Real time operating systems are often used in process control environments, such as manufacturing, refining, telecommunications, and power plants. In these environments, when software makes a request of a remote machine, such as to regulate the flow of a valve, it better happen at the moment it is supposed to or calls are not made, or assembly lines stop running, nuclear reactor control rods don't move into a safe position, or other not so nice things occur.

HISTORICAL MILESTONE!

➢ See how the Cygnus Telecom Server compares to Harry Newton's request to a legacy switch manufacturer to change a single software feature change for his telephone set in 1995 on page 221.

Cygnus's Telecom Server implemented at Maritel, consists of a Intel Pentium based Powerstation system, which has 3 Xircom primary rate interface cards (a potential total of 64 lines), Pika D24 for a 24 channel voice digital signal processors voice resource, and encoding and decoding for voice overs, and the regular networking software and hardware such as Ethernet cards, and TCP/ip software. There is an Amtelco MC1 card for conferencing. There are Amtelco station cards for the operators lines for their headsets providing analog telephone interface for their conversations with the ships.

The database system is Borland's Interbase database, and the operator's workstations are Microsoft Windows NT workstations. The Telecom Server has full automatic caller distribution capability with multiple queues, and tabulation of the reporting data is managed within the Telecom Server. The reporting of the data is generated and displayed on the Windows NT workstations. The management of the system is via a Windows NT system acting as the front end of the system. Management of the call handling, as well as the system, is entirely done by using the computer workstations. There are no dumb telephone handsets, where the operators have to learn unique codes to enable conferencing of the calls. This is completely achieved using the Windows NT interface. The heart and soul of the system is the UNIX real time QNX system.

The advantages of this are quite extraordinary, as Glen Smith states, "In a traditional call center environment, if you had to increase the capability of the call center, you are limited to the feature set within the private branch exchange. With computer telephony integration you can do pretty much what you like... because it is all software written in C++." Although complex programming projects such as automatic call distribution are large projects, other types of software functions can be quickly developed. This allows for faster development of new features for the call center because of the use of object oriented software technology. If the Telecom Server does not have this feature, it is quickly and easily written, tested and

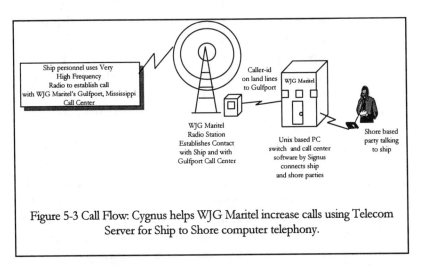

Ship personnel uses Very High Frequency Radio to establish call with WJG Maritel's Gulfport, Mississippi Call Center

Caller-id on land lines to Gulfport

WJG Maritel

WJG Maritel Radio Station Establishes Contact with Ship and with Gulfport Call Center

Unix based PC switch and call center software by Signus connects ship and shore parties

Shore based party talking to ship

Figure 5-3 Call Flow: Cygnus helps WJG Maritel increase calls using Telecom Server for Ship to Shore computer telephony.

implemented. There is not necessarily any new hardware to install, and the new feature can be shipped and installed in minutes to the target system via a modem, or Internet connection. The system can scale to hundreds of operators on a single system. Besides scalability, the system is open enough to be modular, and easily integrated to other systems and business processes.

Call Flow

A call placed from a very high frequency (VHF) radio on a ship is a little different than any other call. A ship will have a designated person or persons who are authorized to place a call. They initiate the call by setting their VHF radio to a prearranged frequency. The crew member will then depress the transmission switch on the radio microphone for 5 seconds without speaking. The nearest land based radio receiver senses this transmission on the frequency, and an automated voice is then transmitted from the radio station to ask the crew member to depress the transmission switch on his radio for another 5 seconds. Once the radio station recognizes the second transmission, it automatically calls the Maritel call center in Gulfport, Mississippi and then connects the crew member to the operator. The operator there will then ask for your ship name and your customer identification, and the number you wish to call. The operator would manually dial the number and using the proprietary telephone set commands, conference the call.

The older system was labor intensive, since the operator had to be in the call cycle most of the time, and had to keep track of the call, and manually update the call records in a database. As the business grew, more and more operators were being needed to manage the traffic. They had 6 operators at a time, working in shifts, 24 hours a day. Since their business was expanding the need to have more operators was also looming. The problem was that a single operator can handle only one call at a time. They were losing calls. Since each call takes a specified amount of time, the manual system could not keep up with the demands. The existing legacy private branch exchange was operating at full capacity to handle these calls, and did not offer any more features to address the more advanced needs of Maritel. It had limited capability of queuing the call, and terminating it to a live operator. Other problems were more human related than computer or telephony related. Once the call had been received from the ship and linked to the land based party, the operator had to remember that the call was in progress. Once there was one call in progress, and the operator had to handle another call, the operator still had to keep track of all the other calls in progress. Once the calls had been completed, the Maritel operator had to manually write the billing information into the database. This allowed for inevitable human error to creep into the system.

Maritel estimated they were losing up to 30% of calls being placed from the ships. Since the manual system created such a bottleneck, these other calls could not talk to the Maritel operator. This lead to some small amount of information being lost. In order to handle more traffic, this system had to be much better at accepting more traffic, by automating the administrative tasks, and servicing customer needs.

In the process of looking for a solution to this, a large private branch exchange manufacturer provided a proposal for $1,500,000, to provide most but not all, of what Maritel wanted. Maritel decided to look into computer telephony integration as a method of providing an alternative lower cost solution to the traditional switching solutions.

Glen Smith said it was important to educate Maritel's management on what was required for appropriate functionality and capability in a computer telephony solution. Key to the success of the project was to solve the immediate issues at the Gulfport call center, rather than deploying computer telephony integration at the radio sites. This laid the foundation at Gulfport that was scaleable to handle the increased volumes. Using other computer telephony technologies in the future, both at Gulfport and at the radio sites, would be easier, and more cost effective.

When an authorized person aboard the ship makes a request, the major difference from the traditional method is the call now terminates on primary rate interface cards on the Cygnus system. This bypasses any need for an advanced legacy private branch exchange. As would any other automatic caller distribution (ACD), the Cygnus system determines the next available operator to take the radio call. The radio portion of the call is answered by the operator in Gulfport. The operator's workstation has a new call ticket form that appears automatically on their screen. The operator enters the ship's name into the form, and if necessary, verifies the person placing the call onboard ship is authorized to place it. The operator transcribes the phone number into the call ticket form, and then initiates an outbound call using an open telephone channel on one of the primary rate interface cards. This is done with a simple click of a button. Once the call is connected, the form automatically minimizes on the screen. The minimized screen becomes an icon, showing the ship name, and lapse time the call has been in progress.

While the call is in progress, the Cygnus system monitors the call for disconnect on the shore side of the call, as well as for any voice energy from the ship side. Once the shore side of the call has disconnected, the billing record for the call is automatically updated. This eliminates an operator's step from the manual keying of this information. If the call is disconnected on the shore side, and there is voice energy from the ship's transmissions, the operator's workstation automatically maximizes the window for that ship, and connects the ship party with the operator, once he or she is available to answer the call. This allows the ship to place another call, or to ask any questions they may have of the operator. Since the operators' calls are being handled in an automated fashion, the wait time in the automatic caller distribution queue is much shorter, allowing a ship party which just completed a call to be quickly presented to the operator. If there is no voice energy on the line from the ship, after the shore party has disconnected, the call billing records are updated, and the icon representing the call on the operator's screen automatically disappears. This is elegant computer telephony integration, which increases profits, saves money, and fosters happy customers.

Glen said there are plans afoot to automate this process even further, allowing the ship to place the call themselves, without operator intervention, to the shore leg of the call. A specialized device attached to the microphone of the very high frequency radio, will automatically identify the ship, and allow them to create touch-tones over the frequency to dial the number themselves.

The key thing to realize is that the automation of the call handling by the operators in a call center manner improves the quality of service to the ships. This allows Maritel to place a large volume of calls, increasing their profits. The system allows the operators to handle more calls efficiently, and easily.

Business Analysis

Maritel uses technology based on more commonly understood, low cost computer technologies, to provide the computer telephony integration and switching. The cost savings have been dramatic, and have allowed Maritel to obtain a competitive edge in the marketplace, as well as providing better customer service and a robust ability to grow in the future.

RETURN ON INVESTMENT

The immediate savings over a comparable solution from a legacy private branch exchange manufacturer was $1,000,000 for the call center solution. Because of the automation now available, Maritel now has over 30% better call throughput and call retention, providing greater revenue for the company. Glen Smith stated, "$500,000 is still a significant investment, as compared to the $1,500,000 proposed by a competitor. If you just wanted to handle calls you could purchase a lower cost regular private branch exchange. The real benefit they derive is the solution is customized to their exact needs, and they receive 30% more functionality than what they would normally could get in the market." Ultimately, the 30% better call throughput using the system will result in 30,000 more minutes per month of billings for the company.

RIPPLE EFFECT

For Maritel's customer's the major benefit is better customer service, call handling, call waiting and more accurate billing of the calls. The function of placing the call from the ship will not have changed, yet they will be better serviced. At the time of writing there were no specific statistics or comments available from the users of the system.

Vital Statistics

The "Shore Connection" - WJG Maritel	
Cost	*Statistics*
Weeks of labor to install system	3 to 4 months with all the software customization needed for the site.
Initial System Cost	$500,000
Est. Ongoing support Cost	not reported
System Size	
# Ports - Total/# Fax.	64 ports for inbound/outbound
# Users	6 operators in call center
# Calls/month	100,000 minutes a month
Computer	Cygnus Telecom Server using a Pentium based Powerstation P6000.
Operating System	QNX, a real time UNIX variant
Industry	*Ship to Shore Telecommunications*
Estimated Return on Investment	2.5 years
Cost Savings	
Manpower equivalency per year	WJG Maritel will not need to hire 6 new operators to handle the increase in call volume minutes
Estimated Manpower savings per year	not reported
Reduction in telephone costs/year	$1,000,000 for initial installation costs.
Earnings	
Generated new income/year	30,000 extra billing minutes/month.
Ripple Effect	
Estimated cost Saving to customer base	Not reported
Vendor	*Cygnus*
Product	Telecom Server
End User	Crew & passengers of Ships
Computer Telephony Technologies Deployed	
PC Switch	Yes

The "Shore Connection" - WJG Maritel

Cost	Statistics
Screen Pop	Yes
Call Conferencing	Yes
Client-Server computing	Yes
Call control via computer workstation	Yes

Table 5-1 The "Shore Connection" - WJG Maritel - Vital Statistics

The "Smart Connection" - Arbinet

T he competition in the long distance marketplace in the United States and Canada has resulted in reduced costs for personal and business calls, and driven competitors to offer voice mail, caller display, and other value added services to find differentiation within the market. For those of us who are fortunate to live in these 2 countries, we should count our blessings There are those in other countries who do not enjoy our freedom of choice for international long distance. Arbinet is a provider of alternative long distance, fax and data services. They work with knowledgeable local companies around the world selling international callback , fax and other services in South America, and Asia. This case history reviews their technology, and shows how a large customer, who shall remain anonymous, has been able to gain larger margins and greater profits by providing these alternative services to multinational accounts. Using an SCO UnixWare platform, Arbinet has been able to provide the scalability and connectivity needed to handle such a task, to manage hundreds of telephone ports and multiple data and voice protocols.

Background

If you live outside the United States or Canada, the telephone company may be the only provider of telephone service and hence its pricing and business policies could be monopolistic. These companies are sometimes controlled by the national government. They control the rates, and eliminate any competition as much as possible. For businesses in those countries it can be very costly to make international calls. As you may have anticipated, many countries do not have the same costs as others to do international calling. This leaves a door open for alternative access providers to allow callers to establish international calls at the lowest rate between the 2 countries. This is called international call back services.

Arbinet started business in 1989 in the voice processing industry. In the early 1990's the opportunity in international callback developed, especially in the area of callback debit card switches. Arbinet has redefined its mission to provide alternate services, of which one of those services is international call back. The alternate access products and services are provided world wide. In order to provide the alternate services Arbinet migrated their switching for debit cards and call back to UNIX based systems on SCO UnixWare.

Arbinet has 2 core platforms. These are called central local nodes, and telephone local nodes. The central local nodes are the large PC switches that are the main gateways which reside in another country. They can handle up to 7 T1 lines or 168 telephone lines in a single personal computer chassis. A telephone local node is essentially an autodialer that resides at the customer's site to aid in the dialing and setting up of calls.

In order to provide the international service, Arbinet selects local partners in those locations to install Arbinet's Central Local Nodes. These central local nodes are UNIX systems, and they provide the voice and data transmission processing. The network is used to move fax, or regular data, and as well to establish international calls using a call back technique. Remote software maintenance by Arbinet is done very easily, since the network makes the equipment as accessible as if the machine was in a room down the hall, rather than on the other side of the world.

Debit Card services in the local market are one of the alternate services. Arbinet provides fax services. They are establishing their own network with their central local nodes operators, so that packets of information, such as fax documents, can be moved from node to node to deliver fax documents internationally, bypassing a 3rd country such as the United States to greatly reduce the transmission costs. Sheila Peterson, Director of International Marketing and Sales said, "So if you are sending a fax, using the Internet from Hong Kong, the cost of sending the fax at the central local nodes, having it moved to the central local node in Paris, France and fax it to there, is nearly free. The customer in Hong Kong would pay a very small fee to have the document sent." Sheila pointed out that they use the Internet as a mover of data, and not for voice calls. Arbinet is not an Internet service provider.

Currently Arbinet has about 1 dozen central local nodes around the world, with the majority in Asia and South America. There are many more committed systems for these markets which will be established in the coming months.

Sheila stated they believe that they are operating, at the time of writing the world's largest PC switch with 8 T-3 lines. A T-3 is also known as a DS3. Each T-3 is equivalent to 28 T1 lines. A T1 consists of 24 telephone channels. Therefore Arbinet has 5376 lines configured in a PC Switch cluster of central local nodes. This is housed in multiple SCO UnixWare systems at their head office in New York city. Alex Mashinksky, president of Arbinet, estimates that by late 1997, they will be operating over 500 T1 lines. The international central local nodes are directly linked via high speed data communication lines to the New York office. This allows them to pass telephone call back routing traffic information between all central local nodes around the world. Moreover, the remote support management of the international central local nodes can be easily done through these links from New York, making, in effect, a virtual PC Switch that spans the globe.

System Overview

The central local nodes overseas provide not only lower cost international call back and enhanced fax services, they also are designed to avoid the local public telephone company from blocking calls going to the system. The methods used to block calls involve screening out information about the call itself that helps the international call back service route the call.

Arbinet complies with local regulations by having regular calls terminate onto a central node. Information is supplied to the central local node to indicate what phone number is calling, and the phone number in the remote country they wish to speak with. This information is usually submitted to the central local node by a transparent local node installed at the corporate office of the subscriber. The transparent local node is a sophisticated dialer that provides this information to the central local node. From a subscriber's point of view, he or she only had to pick up their phone and dial the international number they wanted, a message is played and they are instructed to hang up. The transparent local node calls the central local node, with the international phone number, and the subscribers phone number, and location. This information is captured and signaled from country to country using regular packet switch network technology, using the network of SCO UnixWare based central local nodes. Using routing and rate tables stored on the central local node, the call is set up at the least cost between countries. The subscriber is called back and the party they wanted is on the other line. In this way they are able to set up calls without having to depend upon the local telephone company in the remote country. If one of the SCO UnixWare central local nodes is not functional in the network, the other network nodes act as immediate fall back systems to help set up calls. Because of the service is so constructed Arbinet provides a faster service, for setting up calls, and for transmission of fax data across their international wide area network of nodes.

Call Flow

An example of this would be if you lived in Rio de Janeiro, Brazil and you frequently need to talk to suppliers in the United States. For argument sake, if you make a call from Rio de Janeiro for 10 minutes it costs a small fortune to talk to your supplier in Santa Cruz, California. The cost for your supplier to call you from California is much less for the exact same kind of call. It would not be very practical, if you just sent a quick fax to your supplier say "CALL ME".

An international call back provider would have a system in the United States. You call that special number. Then you enter your access code for billing purposes, and the number of the party you wish to talk to in the United States. You hang up. This takes all of a few seconds on the call from Rio de Janeiro. The system in the United States then phones you back while at the exact same time, tries to phone the number you have supplied in the previous call. It then conferences the two calls together to make them one call. You are billed at the rates that a customer from in the United States would be charged to place a call to Rio de Janeiro from Santa Cruz California, plus any US based long distance charges that may have been incurred.

This has been a very fast growing industry and the ability to service and support these systems from afar is important. The SCO UnixWare systems make that an easy job via the networking facilities within the operating system.

Business Analysis

In this case history both the anonymous reseller of Arbinet alternate services and the multinational companies and individuals greatly benefit monetarily. Sheila Peterson said this portion of the long distance market is growing 100% per year.

RETURN ON INVESTMENT

The major benefit for the anonymous reseller of Arbinet's services is that they are now able to sell at a much greater margins of profit than they could in the very competitive long distance market in the United States of America. Even with the competitive discounts against the foreign countries public telephone network provider, the reseller still makes 30% more margins. In comparison, within the United States, wholesale long distance rates are down to 2.5 cents a minute, and the U.S. Government is purchasing bulk rates at 1.9 cents a minute.[58] This is forcing resellers in the US market to provide value added services in order to increase profitability, while competing in a relatively new market in international dial back, at much higher margins, with less need to provide alternative services.

RIPPLE EFFECT

The companies that are doing international business or are organized as a multinational, also see the reduction in costs and increased communication with their clients, and their ability to service the needs of their customers. When communications costs drop, generally the profits of the company increase, since it is easier to service existing customers, and increase the opportunities to search for new prospects.

Generally speaking the end user customers see at least 30% reduction on the costs of placing international calls as compared to the rates of the local telephone company.

Future International Callback Market implications of PC Switch Technology

Alex Mashinsky provided some insight as to the market forces at play which will change the international callback market. The PC Switch technology provides an important fundamental catalyst to this change. This fundamental change will effect the overall private branch exchange market. The PC Switch is forecasted to be as revolutionary as the advent of the personal computer was to the oligarchy of computer vendors of the early 1980's. The same revolution is taking place all over again with the advent of reliable, scalable, and standards based PC Switch technology.

Another factor in the international Telecommunications market, is that the organization and reconciliation of accounts of long distance charges between countries is going to dramatically change. Currently, the International Telecommunications Union Telecommunications ITU-T, is a United Nations organization. Its current membership consists of 170 nations. Its primary goal is to set standards and to help regulate frequencies. Alex explained the ITU-T members reconcile their international long distance accounts once every quarter.

Because of the low cost, open standards based PC Switch technology, this will allow for many new lower cost value added technologies and services. The cost of entry into the market using PC Switch technology, will be the impetus for countries with monopolistic long distance services to open their markets to competition within their own markets in the next 2 to 3 years. Alex estimates such changes will create up to 40,000 new vendors of long distance telephony services worldwide. A dramatic increase from the existing 170! The ITU-T can manage the 170 members, but managing 40,000 is a different matter. One of the major challenges is that the billing systems will have to reconcile accounts in almost, or in real time on a global basis. This is a challenge that Arbinet is working on today. Alex firmly believes this is the birth of a multi-billion dollar industry, which will grow to be very large in the future.

Vital Statistics

The "Smart Connection" - Arbinet	
Cost	*Statistics*
Weeks of labor to install system	3 weeks to deliver customized Central Local Node. Some delays occur when obtaining lines in foreign countries.
Initial System Cost	Central Local Node with one T1 is about $28,000[59]
Est. Ongoing support Cost	
System Size	
# Ports - Total/# Fax.	8 T3 (a.k.a DS3). 1 T3 is equal to 28 T1s. Each T1 has 24 lines. Total number lines at the time of writing (autumn 1996) 5376 telephone ports! They claim it is the largest PC Switch in the world.
# Users	"Tens of thousands of users"
# Calls/month	"Millions of minutes per month"
Computer	Pentium based industrial based personal computer.
Operating System	SCO UnixWare
Industry	
Estimated Return on Investment	
Cost Savings	
Manpower equivalency per year	not applicable
Estimated Manpower savings per year	not applicable
Reduction in telephone costs/year	30% reduction as compared to local telephone rates in foreign countries.
Earnings	

The "Smart Connection" - Arbinet

Cost	Statistics
Generated new income/year	30% increase in gross margin as compared to the local long distance market in the United States.
Ripple Effect	
Estimated cost Saving to customer base	30% reduction in cost for voice and fax calls.
Vendor	
Product	VoiceNet
End User	Anonymous (we are not allowed to tell you.)
Computer Telephony Technologies Deployed	
PC Switch	Yes
Internet based Fax Solutions	Yes
Client-Server based computer telephony	Yes
Remote Worldwide System Management via UNIX open standards	Yes

Table 5-2 The "Smart Connection" - Arbinet - Vital Statistics

---

[58] These rates are high volume bulk wholesale rates which large corporations in the United States of America can contract with the long distance companies. The values were made public at the Computer Telephony Demonstration Fall 1996, by Harry Newton on October 31, 1996 in Orlando Florida

[59] Teleconnect Magazine September 1995 in the article "International Callback Roundup" VoiceSmart section of the roundup. Published by Flatiron Publishing Inc. 12 West 21 Street, New York, New York 10010.

The "Intranet Long Distance" - AlphaNet's Mondial Network

In a world where computer and communications technology rapidly changes it is very easy to be caught up in the whirlwind and excitement of new technology. Technology can be only become an asset when it is used to solve real business problems.

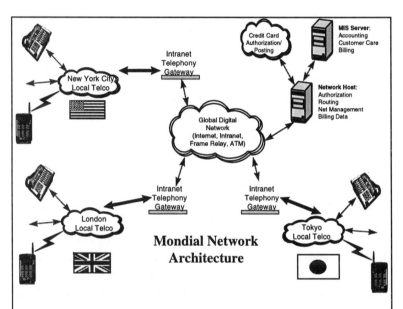

Figure 5-4 AlphaNet's Mondial Network drastically reduces the cost of international long distance voice communications using SCO UnixWare and AlphaNet's patented technology for Intranet Telephony Gateways.

Every so often a new technology gem shines through the technology storm and fundamentally changes the rules. This makes a significant impact on the bottom line. AlphaNet's Mondial Network is one of those cost reducing technologies. Mondial is the first worldwide based long distance service that uses TCP/ip technology using frame relay transmission to transmit voice, using an Intranet of SCO UnixWare servers.[60]

Background

AlphaNet Telecom[61], Toronto, Canada, is a developer of UNIX based fax, and Intranet based voice communications. Historically, the company has provided their InnFax products to the international hospitality industry and with their "follow-me fax products" for businesses around the world. AlphaNet is also a telecommunications wholesaler with their Mondial Network.

Mondial technology is significant for three reasons.

1. It dramatically cut costs of voice communications around the world.

2. It also marks a unique point in the history of the convergence of communications and computer technology. Both are now possible in the same PC Switch architecture. This dramatically reduces the cost of deployment.

3. It will also allow for the growth of many new solutions in the near future, since corporate data and voice are encapsulated and manipulated in a low cost open standards base technology.

The savings are both in the technology itself, as well as the method of transmission. Instead of transmitting over international tariff voice lines, the cost of transmission is restricted to the fixed underlying costs of a frame relay data and asynchronous transmit mode networks. In some countries, the international long distance rates are subject to the whims of the local telephone monopoly of a particular country. Using a very high speed frame relay network bypasses a country's international call rate structure. Long distance calls can also be placed within the same country as well, and still be able to provide customers with a significant saving for the same reasons. How this achieved is discussed below.

In September 1996, AlphaNet launched their international Mondial demonstration telephone voice network. On January 13, 1997, major steps were taken to lay the foundations of the commercial version of the Mondial Network. An agreement with Scitor allows AlphaNet worldwide access to frame relay (45 megabits/second) and asynchronous transfer mode (622 megabits/second) circuits, facilities management, network monitoring, and performance management and support.   Scitor is a world leader in telecommunications infrastructure.   AlphaNet will have at first a point of presence in 15 countries when the commercial network is launched, in early 1997, with 20 more countries shortly thereafter.

Each point of presence is a AlphaNet Mondial computer that has both regular voice card technology to interface with the local regular telephone network, as well as, AlphaNet's own software technology that moves that voice over its own network to another point of presence in a foreign country, where the process is then reversed.   As a user of the Mondial network, you place a call as you would with any other 3rd party long distance service. You call an access number, wait for an interactive voice response application to collect your access number, and the number you want to dial. The call is placed normally.

System Overview
Before a detail review of system overview can begin, we need to understand a little how the Internet works, and why it was important for AlphaNet to use an Intranet network typology. For their Mondial Network.

You may have heard of, or have used, personal computer based voice over the Internet products, such as Internet Phone. In the public domain, software products like CuSeeMe, has been ported to different operating systems including UNIX.   These products allow a user to use their computer system, and a microphone and sound card with speakers, to engage in a telephone like conversation. With some products, like CuSeeMe , and Microsoft's NetMeeting, they allow you to use a small video camera attached to the computer peripheral to do video conferencing. The main advantage of this type of conversation is the very low transmission cost to another person in the world. The software is free, or commercial versions are priced like low cost utility software. There are no toll charges to talk to a friends and family across the country or 4 continents away. The only cost is the low monthly flat fee for access to the Internet from the local Internet service provider.

There are technical challenges to Internet based voice communications.

A computer on the Internet needs to know the Internet Protocol address (i.e like a telephone number) of the other computer on the Internet, in order to send data to that other computer. Whenever you reference a Internet site such as *sco.com*, *sun.com*, *ibm.com*, or *scheduledsolutions.on.ca*, the site name is cross referenced by a special Internet server called a Domain Name Server. The Domain Name server sends back to your Internet application the Internet Protocol address. The major central clearing house for Internet Protocol addresses is an organization called the InterNIC. Its main purpose is to assign Internet Protocol addresses for organizations.

Although the Internet Protocol addressing scheme was designed to support over 1 billion Internet Protocol addresses, very large blocks of addresses have been reserved for years by large entities such as the US military, universities, and large corporations. With the large growth in the Internet, the remaining blocks of numbers are in danger of running out. (Currently, there is a revised addressing standard under review to solve this problem.) Once an Internet protocol address is assigned by the InterNIC, the registration is then propagated around the world to different computers. Remember, the Internet is designed to decentralize information and command and control structures.

Once the Internet protocol address (or "ip address" - pronounced "eye pea address") has been registered to a company or individual with the InterNIC, then the Internet protocol address can then be associated with a domain. A domain is a logical organization of types of Internet addresses. Examples of domains are, *.com* for commercial enterprise sites, or *.ca* for Canadian domain, or *.gov* for US government sites. My company, Scheduled Solutions has a site name of *scheduledsolutions.on.ca* . This means there is a site called scheduledsolutions that is registered in the Canadian Domain, within the province of Ontario.

It is up to a system administrator and management to make the decision under which domain they wish to register their site. Once this decision is made, it effects how the rest of the world can access you through this important medium. After a site name is registered under the domain, then this information is propagated to many thousands of domain name servers around the world.

Conceptually, domain name servers are analogous to electronic telephone directory services. If you wanted to have a terminal session with a site called uunet.ca using a Telnet program, or use your world wide web browser to look at www.whitehouse.gov, a domain name server is consulted, and the Internet protocol address is then retrieved for the site. The Internet protocol address is then used by your Internet software. The Internet protocol address for a domain name server at the major Canadian Internet service provider, UUNET Canada (uunet.ca), is 142.77.1.1 . Internet service providers are analogous to a local telephone companies. They provide access to the Internet for their customers, as well as manage the UNIX systems that maintain the connections and Internet applications their customers need. Domain name servers allows users around the world to not have to remember the unique Internet protocol address. They only have to refer to the site by its domain name.

Because of the limits of the current Internet protocol addressing scheme, the InterNIC has given assigned block of Internet address to the Internet service providers. Instead of each individual having a unique Internet address like their own phone number of their home or business, the Internet Service Provider will randomly

> Wonder why the cost of the Internet is low? One factor has to do with countless hours of free time of very skilled dedicated individuals. There is a volunteer who spends about 4 hours every day to maintain the Canadian domain! This is not atypical. It is because of the unsung dedication of volunteer system administrators and other such as these that has made the Internet what it is today.

assign one of their currently unused Internet addresses, within their assigned block, when an individual computer system has established contact with their site. For the duration of that session of being logged onto the Internet, the individual computer will have a unique Internet protocol address. After they disengage from the Internet, the Internet service provider will then allocate that unused Internet address to another customer as soon as the next customer logs onto their system.

Herein, lies the problem with using products such as Internet Phone, in that you do not know the Internet protocol address that has been assigned to the person that you want to talk to, until both parties of the conversation have successfully connected to the Internet. In order to set up a call, a prearranged time of the call has to be established by either e-mail, or by a regular long distance telephone call. Once both parties are connected to the Internet, they have to exchange email, or a long distance call to inform each other of their newly assigned Internet addresses. Then they can use their voice over the Internet software to talk to each other. Some companies such as Microsoft are setting up special directories on the Internet to allow users to search for their prospective parties, without having all this hassle. But the problem will not easily go away, until a newer standard of Internet addressing has been rolled out.

The quality of sound of voice on the Internet software has been plagued by the very nature of the Internet itself. The Internet was designed, so in the case of a nuclear cataclysm, communications could still be obtained with the other sites on the Internet. This is possible on this network, since the routing information is not centralized through a single computer.

The Transmission Control Protocol/Internet Protocol (TCP/ip) software manages the routing and bundling of information into packets. Each single email message, sound, voice, and data record is broken down into individual packets and then sent to the Internet address specified in its destination. You may think that the information between a computer system in Yellowknife, North West Territories, Canada to a computer system in Sydney, Australia would go directly from one system to the other. This is rarely the case. The information will hop from one computer to another system in a long chain of computers and network routers before reaching Sydney, Australia. If one of the computers in the chain becomes unavailable for any reason, the packet stream from Yellowknife, then is routed via another route, let's say, to a computer system in Johannesburg. The routing may be through a mixture of high and low speed telecommunication lines. This then causes delays, and more importantly breaks up the sound that one may hear on the conversation.

All these factors come into play, in providing reasonable, (i.e. not great) sound quality, and relatively low cost access to voice communications over the Internet.

Would you put up with the cost of having all this computer hardware and software, and training, needed to make a call? Does it really make sense to have to use a $1,500 to $2,000 computer hardware investment with a $100 software package, with many hours of training and fiddling to make it work, in order to save $50 on an international telephone call? Probably not, unless you are planning to do this very often, and for a long time.

If you were a business, would you trust your next multimillion dollar deal with this kind of communication? Probably not. You would pay the long distance charges to make sure that you would be able to make the connection any time you wanted, without interruption, and with the quality you expected.

The Mondial Network solves all these issues in a single solution. It provides both the tremendous cost reductions of the Internet based open standards technology, and the reliability, plus ease of using the regular telephone service. These costs are passed on to the general public through AlphaNet's customers. As Al Gordon, Chairman and Chief of Technology of AlphaNet, said "AlphaNet is a 'carrier's carrier'. In other words, we will be an international carrier to existing aggregators, such as cellular operators, call-back providers, carriers seeking evaporation of international traffic, calling card providers, etc." In order for businesses and individuals to use the Mondial network, Al Gordon said, "We (AlphaNet) will have an end-user service offered through Regional Marketing Partners (RMPs) in various countries."

Call Flow

With the Mondial solution you do not need to be a computer 'guru' to place computer based calls using the Internet. In fact, you do not need a computer at all. Mondial allows you to use your regular telephone to dial into one of the local points of presence, and using an interactive voice response system enter in your account number, password and dial the long distance number. The call is then connected to another computer (also referred to as a "*point of presence*" or POP) in the country you are calling. The computer in the other country is sent the telephone number you want to speak with. The remote computer, using regular computer telephony board hardware, dials your desired number. The remote computer and the local computer establishes the connection between your telephone with the dial tone in the other country. Between the two computers, voice is sent in TCP/ip packets, and reassembled into regular voice for transmission over the local telephone lines in the remote country or location. The two computers are connected using an "Intranet", or private network, that are communicating using the TCP/ip network software. There is a benefit of low cost leased communications link between the two systems. It is a SCO UnixWare computer system that is directly connected, using leased high speed data lines.

The Mondial demonstration network, which existed until January 14, 1997, had points of presence in Australia, Hong Kong, Japan, Canada, Great Britain, and United States. The demonstration network was ended, in order for the commercial network to be implemented.

Scheduled Solutions was given access to this network in order to research this book at a greatly reduced cost, as it would have been for the cost of the regular long distance carrier. The overall quality of the phone calls is very acceptable. Using the network to perform taped interviews for the case histories, there were no major downgrades in the performance or the quality of the network. This afforded a chance to review the material many times before writing the case histories. The tapes showed the quality of the sound was constant, and there was very little interference.

Business Analysis

The essential difference is the nature of the transport of the call has dramatically reduced the cost of research and production of this book. In a production environment AlphaNet is forecasting at least 50% reduction on the best rates from leading providers of long distance services, such as Bell Canada, Sprint, or AT&T Canada. The greatest impact will be the transcontinental traffic, where the largest margins in the business of long distance still are a major revenue source for the competition.

RETURN ON INVESTMENT

The best measure in this case history, is the fact that AlphaNet has provided the service at the lowest possible cost to Scheduled Solutions. Many of the details of the original research in the case histories that you see in this book are a testimonial to the quality of the service, and as well as the support by AlphaNet for this book.

For a business, the technology will dramatically reduce the costs of doing international business and long distance based business.

RIPPLE EFFECT

The ripple effect in this case history is directly to you, the reader of this book. The information contained in this book, has been garnered with the help of AlphaNet Mondial Network.

Vital Statistics

The "Intranet Long Distance" - AlphaNet's Mondial Network

Cost	Statistics
Weeks of labor to install system	None - just dial normally
Initial System Cost	No initial cost to subscriber other than regular telephone line costs
Est. Ongoing support Cost	None

System Size

The "Intranet Long Distance" - AlphaNet's Mondial Network

Cost	Statistics
# Ports - Total/# Fax.	
# Users	
# Calls/month	
Computer	
Operating System	SCO UnixWare
Industry	*Telecommunications*
Estimated Return on Investment	
Cost Savings	*At least 50% lower than Published long distance rates.*
Manpower equivalency per year	not applicable
Estimated Manpower savings per year	not applicable
Reduction in telephone costs/year	very significant
Earnings	
Generated new income/year	not applicable
Ripple Effect	
Estimated Saving/Benefit to customer base	lower production costs
Vendor	*AlphaNet*
Product	Mondial Network
End User	Scheduled Solutions
Computer Telephony Technologies Deployed	
PC Switch	Yes
Internet based protocols	Yes
Client-Server computing	Yes
Remote Worldwide System management via UNIX workstation	Yes

Table 5-3 The "Intranet Long Distance" - AlphaNet's Mondial Network - Vital Statistics

---

[60] Author's Note: Since this case history was originally written, AlphaNet Telecom Inc. has moved to use its own encoding scheme on frame relay to move the voice traffic as data without using TCP/ip as the method of packaging the voice. TCP/ip is still used between the points of presence for traffic control and other administrative purposes. Although TCP/ip to encapsulate the voice as data does work, it added overhead which consumed bandwidth, which could be used to sell more minutes of long distance. In addtion, a "private" frame relay backbone provides the type of reliable low-latency transport required for toll quality voice.

[61] *The author wishes to express sincere thanks and appreciation of Alastair Gordon, Chairman and Chief Technology Officer, AlphaNet Telecom, for allowing Scheduled Solutions to use the of the Mondial demonstration network for research for the book. AlphaNet's support of this book project is deeply appreciated.*

Chapter 6

Integrated Solutions

This chapter reviews HOW to implement an integrated computer telephony UNIX solution

In this chapter, we will present a business case, and an request for proposal for a **fictitious company**. Although the company name is false, the business case is real, and has been altered to present a true multiple branch international company that has the same problem - sharing information, and providing timely customer service.

Figure 6-1 Careful planning will ensure your integrated UNIX computer telephony solution is a success. Reprinted with special permission of King Features Syndicate.

In the previous chapters we have shown why where and how computer telephony has been used by other companies have used different kinds computer telephony applications to save time and money for themselves and for their customers. In this chapter, the goal is provide a business situation that will require all computer telephony technologies to come together to service the external customer service and support, and as well as providing service for the internal staff of the multinational corporation. This chapter will show how the products are integrated together to make a total enterprise solution. This may be more technical than is required for a business manager, but it will highlight what is important.

Pearls of Wisdom for successful integration - Genesys

The following case history diverges a little from the others in this book, in that it focuses on the actions needed, by all parties, to successfully install and roll out a computer telephony integration project. In this case history, the project was for Charles Schwab, and it was performed by Genesys Telecommunications Laboratories, using their T-Server product on the Sun Microsystems platform. Charles Schwab required inbound call handling, with screen pop technology for about 1600 customer service representatives at 4 call centers averaging 400 customer service representatives located across the United States. The interviews with Bill Mitchell, Implementation Manager and Gary Lam, Marketing Communications Manager of Genesys located in San Francisco, California, provided some very interesting pearls of wisdom about the issues that need to be addressed by all parties involved, for a computer telephony integration project to succeed. These pearls of wisdom are applicable to a large project such as Charles Schwab, as well as for smaller computer telephony integration projects for small workgroups.

◄ For further information on Genesys T-Server and other middeware products, please refer to page 206.

Background

Charles Schwab used T-Server and integrated it with an application they had already written. Genesys T-Server populated the agent's screen fields for the screen pop. The database system that Charles Schwab integrated was a legacy mainframe system used for customer service, and account maintenance for tracking payments.

The inbound caller would reply to prompts from an interactive voice response application, which would then communicate the customer information with an Aspect ACD queue. The call was integrated with the Charles Schwab application via Genesys T-Server, on a Sun Microsystems Sparc system running Solaris 2.4. The purpose of the T-Server is to act as a translator between the telephony world and the data network world. In this case the switch was an Aspect, and all telephone networks such as T1s terminated on that system. Using the Aspect Link bridge product, the Aspect switch communicated with the Sun Sparc system running the T-Server software. In turn, the T-Server communicated with the legacy mainframe system to coordinate the screen pop, screen transfer, or screen conference.

The system took 4 weeks to install all 4 call centers. Each call center took 2 days to install, and then 3 days of testing by Genesys and Charles Schwab staff. The four call center sites are located in Colorado, Florida, Arizona and another in Indiana.

Pearls of Wisdom

The following is a discussion for managers who need to know what things to look for, and to do, when embarking on their first computer telephony integration project planning and implementation.

I.   First, you have to take stock of what you have. You must identify the key elements, key departments and key personnel necessary to outline the computer network topology, telephone switching topology, and telephone switching call flow. Make sure all the team members in house know the technology and the capacities of your switch, call flow, LAN system and that LAN and telephony administrators understand the daily running characteristics of the load.

II.   Bill's experience has shown that some of the hold ups in an implementation are caused by not having the right players on the team. What he has seen in the past is teams that are not fully knowledgeable tend to trip over things that were unexpected. They did not take stock of what they had in house. This delays delivery of the solution that they had already promised to upper management.

III.   Once you have the right team members onboard you can write a functional specification of how the solution will be put together. The functional specification includes a detail of the network, and telephony, data topologies and call flows. It consists of telephony, computer, and call flow diagrams, and outlines of the scope of the work to be performed, and what the architecture is, from a data and telephone network points of view. With a good specification that shows in detail how the solution is to be put together, it can be delivered to a system integrator to develop.

IV.   It is not necessary to cross-train the telephony and computer specialists in each others trade for successful implementation and integration. It is important that each of these team members understand the impact of their systems on each other's systems. For example, it is important the telephone specialist understands the impact of the messages coming across the telephony link from the switch to the computer system, on the data network. How these messages are used by the receiving computer system and when these messages are needed is important to comprehend. From a troubleshooting perspective, both parts of the team need to be in sync and in tune with each other's systems.

V.   Once the team is assembled, and once the functional specification is done, it is important to set up a triage[62] team to deal with any problems that present themselves. They should list the most likely problems, and have plans ready to deal with those scenarios. This allows for quick response so that they can "have all hands on deck to troubleshoot this problem", said Bill. The triage teams should consist of the telephony, network and computer specialist, and a business manager who can make decisions in context of the needs of the end user.

VI.   "Without a shadow of a doubt", said Bill, "a prototype, and proof of concept should be done. There is usually a lot of pressure from upper management to complete these projects". Prototyping is the best means to ensure on time and budget completion. A *scaleable* (UNIX systems do this very well) proof of concept will identify the issues and problems and allow solutions to be more easily worked out. It gives the team the ability to go to senior management with believable costs and implementation schedules that are realistic and deliverable. Management may also have the option to commit more resources to accelerate the delivery of the project.

A.   A pilot project prompts upper management to do two things:

    1.   Ask the appropriate questions and to allocate the proper amount of resources and dollars

    2.   It allows for a checkpoint by upper management to ensure the implementation team has the proper staffing requirements or give them the go ahead to secure the needed assistance, such as bringing in consultants, and other resource requirements to do the job properly and on time.

B.   The Pilot project will also provide an impact study for the system administrators of the LAN, so they can anticipate potential problems with the LAN or WAN requirements.

VII.   "One of the things that worked so well at Charles Schwab was the preparedness they took with the implementation and triage teams. Bill said "I've worked on other projects with other companies (which were not using Genesys products) where the different team members are 'ripping' at each other because the implementation was not going well. This was not the case with Charles Schwab. They did a very good job of identifying what team members needed to be there, and assembling the triage teams to jump on problems very quickly. Charles Schwab worked very closely with Genesys to partner and understand the technology. The team assembly, and the pilot, and cross knowledge of team members were the key reasons why it only took 11 days to really implement such a large solution for Charles Schwab."

VIII. The importance of the selection of team members for implementation and triage teams and cross knowledge of the telephony, and computer technology resulted in a very interesting side benefit for Genesys. Charles Schwab had taken such a proactive project management approach, and took pains to learn and understand the Genesys installation, that although they had contracted Genesys to be onsite for 3 days to install, they sent the Genesys staff member home 1 day early on the first week. For the remaining 3 weeks, the Charles Schwab staff installed the Genesys solution on their own, with remote telephone support of the installations in the remaining 3 cities. This benefited Charles Schwab because they became immersed in the CTI solution, and as well, saved Genesys some resources.

IX. In the case of Charles Schwab, where they integrated their own technology and interface to the T-Server technology or other API, it was very important for the development team of the end user customer to work closely in developing the interface specification, early in the project. They needed to understand and test the integration between these systems and do it successfully. The sooner the end user developers can get the interface developed and tested, the sooner the proof of concept prototype can go forward.

X. Charles Schwab came to Genesys requesting them to help educate Charles Schwab's developers about the computer telephony interface, and to understand the issues associated with telephony event programming. It is not a common skill to have programming staff that understand telephony events. Charles Schwab sent their developers to Genesys or worked remotely with them so that Charles Schwab developers could learn from Genesys. Genesys engineers learned to understand the needs of the Charles Schwab functionality and how it works.

XI.   An important function of the manager of the implementation and triage teams, is to really be able to assess the skill sets that comprise that team. Making sure that each team member is competent in their job function, (this may require some training), is very important. Computer telephony integration applications tend to be mission critical, and being able to deal with a problem as soon as it arises makes for a better team.

XII.   Gary Lam also pointed out that call center staff training is an important factor for successful cut over of a system. Ensuring that the supervisors are trained in advance, with sufficient warning, and providing appropriate training on the software, and call handling of the staff will go a long way to reducing the startup costs and increasing the long term success of the call center.

Business Analysis

Although the focus of this case history is on the managerial issues, there is nonetheless a powerful business story about how Charles Schwab has benefited from computer telephony integration. Here are the basic facts:

1. There are over 1600 agents

2. They take over 300,000 calls per trading day of 8 hours, which operates 51 weeks of the year.

3. The staff that take these calls are stock brokers who hold a Series 7 certificate to allow them to trade securities. The average industry salary is about 50,000 per year. Please note this is an industry average, and not necessarily the salary paid at Charles Schwab.

4. Computer telephony integration saved an average of 20 seconds per call.

5. The average cost of 800 number minutes purchased in large volume is about 5 cents per minute.

All these facts come together to tell some pretty interesting savings.

RETURN ON INVESTMENT

The return on investment is hard to determine because the cost of the installation is not known. One can calculate some impressive numbers as to the manpower and salary savings the system is providing. The computer telephony integration has gained on average about 13% more productivity for each stock broker. With 1,600 such agents, this translates to an equivalent of 208 more qualified staff. This has allowed Charles Schwab to handle more customers, be able to offer more services and build customer loyalty/satisfaction by providing more efficient services. It is an additional $10,416,667 worth of manpower productivity. This is money that is already being spent, however the money is going to much better use.

As for telecommunication costs, computer telephony integration has provided an approximately $1,275,000 in savings in 800 number minutes.

RIPPLE EFFECT

For Charles Schwab's customers the computer telephony functionality has eliminated 53,125 days (each day is 8 hours) of hold time. This means that customers are able to process their transaction requests much faster.

Vital Statistics[63]

Pearls of Wisdom for successful integration - Genesys	
Cost	*Statistics*
Weeks of labor to install system	1 week in Genesys labor over a 4 week period in concert with Charles Schwab staff
Initial System Cost	not reported
Est. Ongoing support Cost	not reported
System Size	

Pearls of Wisdom for successful integration - Genesys

Cost	Statistics
# Ports - Total/# Fax.	Not reported
# Users	1600
# Calls/month	6,300,000
Computer	Sun Sparc stations
Operating System	Solaris 2.x
Industry	*Financial - Investment*
Estimated Return on Investment	not reported
Cost Savings	
Manpower equivalency per year	equivalent of 208 brokers
Estimated Manpower savings per year	$10,416,667 assuming industry modal salary of 50,000./year
Reduction in telephone costs/year	$1,275,000 in 800 minutes assuming $.05 cents per minute
Earnings	
Generated new income/year	not reported
Ripple Effect	
Estimated cost Saving to customer base	53,125 days per year (8 hour days of on hold time)
Vendor	*Genesys Labs*
Product	T-Server and others
End User	Charles Schwab
Computer Telephony Technologies Deployed	
Interactive Voice Response	Yes
Middleware	Yes

Table 6-1 Pearls of Wisdom for successful integration - Genesys - Vital Statistics

---

[62] "Triage team" is a generic term, and not specifically referred as the same at Charles Schwab.

[63] Calculations done by the author using information provided by Genesys Labs and other sources. The 800 minute rate can range from 5 cents to 10 cents a minute depending on volume, interstate rates etc. I used 5 cents per minute as "average",

and as well a more conservative number to calculate money saved. Numbers provided are based upon industry standard knowledge rather than upon direct quotes from Charles Schwab.

Overview of the Fictitious Minden Biotech RFP

The purpose of the fictitious Minden Biotech Request for Proposal (RFP) is to demonstrate how to integrate a computer telephony solution for a multiple site enterprise, using UNIX based products. This overview is to help the reader understand the main issues that are to be addressed by the respondent - Capri Systems Inc. For full details and copy of the RFP please see

The Fictitious Minden Biotech Request for Proposal on page 347.

Minden Biotech is a biotechnology company that has its main line of business in biotechnological testing of agricultural commodity products, and has also acquired two companies that provide medical laboratory results for the insurance industry. The insurance medical laboratories are in Farmington, New Jersey, while the another is in London, United Kingdom. These were two independent companies, and now are operating divisions of Minden Biotech. Each location has its own independent computer and telephony resources that have evolved over time. The RFP has specifics about how the London office is allowed, by regulatory constraints, to report test results outside the United Kingdom. However, the management of the company realizes that to grow the company, it needs to modernize its computer and communications infrastructure. This is to facilitate the development of communications within the company, as well as with the customers, and the public at large. Since the main product of the company is confidential information produced from its laboratories, the demand for its timely and accurate dissemination is of prime importance. Moreover, the ability to do this efficiently and more aggressively in the market will provide the company with the potential of a sustainable competitive edge.

There are two phases in the RFP. Phase I is to rebuild the communications and computer infrastructures, using any existing investments when they are deemed still valuable for the long term. This includes the implementation of a modified "off-the-shelf" Informix database application product called InGenetic. (*The name of the product is fake*). This is the main application for the corporation. It has been designed for this industry for managing and reporting from receipt of samples, and laboratory work up, and to provide its reporting functions, and complete accounting. It is being customized by the MIS department of Minden Biotech for all the locations of the company. It is an enterprise wide system, which connects into other applications, such as desktop applications. It is this application which the computer telephony functionality is to be built around. The goal is to build the computer telephony functionality and voice integration capability within three offices with call center capabilities.

Phase II of the project is to migrate the Phase I: solution to be a single virtual call center and reporting system for all inquires into the company. This is for sales, laboratory production, and test results. There are some jurisdictional restrictions on what kind of information and to where it can be sent.

Major Business Issues
The major business issues for this RFP are the following:

1. The overall goal is to harmonize the disparate systems that are currently used to manage the business, and to increase the ability to communicate within the company, as well as with customers. Because of the merger and acquisition of the different businesses that now make up Minden Biotech, this has become an important infrastructure and customer relationship issue.

2. The ultimate goal is to allow customers to have a single phone number to access any of the divisions of the company, within regulatory limitations.

3. Flexibility of customer response - In the latter half of the solution, the ability for any one call center in the United States offices to cover for each other, will allow for the greatest amount of customer response. For the UK office, this will also be required, as its operations are expected to expand into continental Europe in the near future.

Major Computer Telephony applications that are addressed

The major computer telephony applications that are addressed in this RFP are:

1. Voice messaging system with integrated fax, voice and email delivery.

2. Interactive Voice Response for test result handling, and call routing.

3. Facsimile Server to fax out test results

4. Call Center Solution with "screen pop", "screen transfers", and predictive dialing capabilities to handle all inbound calls for order processing, and outbound test results reporting and marketing.

5. Internet and Intranet integration including email and Web based services.

These major computer telephony applications are to work in concert with each other, so as to be a single cohesive system working with the selected database solution. The computer telephony applications are not isolated from the corporate environment, but tightly integrated to other services such as email and Internet/Intranet browser access.

Usage and relevance of the RFP Response

A cautionary word or two is probably in order. First, although the RFP is based upon a mixture of real life business challenges and opportunities, the business problems being solved are based upon some imaginary situations. Hence, the response to the RFP is realistic to the issues contained therein, but not necessary adaptable to every company and every situation in real companies. Please use the following reply as an educational tool, rather than being a solution that is laid out in stone to use in your company. This is meant to show how some typical challenges are met with specific software tools. The respondent's solution is the solution for this imaginary company, but the knowledge you shall gain as the reader is valuable and current at the time of writing of this edition. It is important to realize that the response by Capri Systems is one of many possible kinds of recommendations for the business problem. The purpose of publishing Capri Systems response to the RFP is to show how the technology would be integrated, rather than declaring a winner. The winner is you, the reader, in being provided with insight into how the solution is put together.

Capri Systems Inc. Response to the Minden Biotech Request For Proposal[64]

Figure 6-2 Capri Systems provides a response to the Minden Biotech RFP using Sun Microsystems based solutions.

This response to the fictitious RFP is supplied by Capri Systems of San Jose California. The challenge faced in this RFP in that there are many different ways to solve this problem with a different set of UNIX based computer telephony integrated solutions. The purpose of this response is really more to educate the readership of this book, rather than to sell a particular solution.

Executive Overview
CHOOSING THE RIGHT CALL CENTER SOLUTION

Your call center solution has become a productivity tool, empowering your business with an array of Customer management capabilities well beyond simple call answering and product fulfillment. Capri Systems has a solution designed to handle the customer satisfaction demands of Minden Biotech today - as well as those of the future.

PARTNERING FOR SUCCESS

Businesses nationwide have discovered the benefits of partnering with Capri Systems to provide their short and long term customer management needs. Choosing Capri Systems means that you have the creativity, business expertise, and superior support record and the technological innovation to help you exceed your organization's objectives.

THIS IS WHAT PROVIDING A TOTAL SOLUTION MEANS TO US

Capri Systems ensures that your business can take advantage of our total Customer management capabilities. In other words, we have the ability to help assess your current situation and future needs. We will apply that knowledge to what we know and understand about today's dynamic business environment. We consider your desire to control costs, increase employee productivity and improve overall communications efficiency. We design our solutions to thoroughly suit your needs, then integrate it seamlessly into your unique business environment. We train you and your employees to effectively use all of the capabilities that we provide. Most important of all, we are always available to help you in a crisis as well as a routine situation. We provide system expertise, superior products and services, industry knowledge, and absolute commitment to ensure that your organization succeeds, as well.

The Major Application Components
Capri Systems solution includes the following key components:

- Protect as much existing equipment as possible
- Provide as much transparency as possible between locations
- Provide uniform voice messaging across PBX platforms
- Allow Computer Telephony Integration applications to be added as needed
- Ensure a high level of reliability in key components
- Recognize which vendors best handle which components

Our solution includes:

- Integrating to the existing Siemens, NEC, and Harris switches
- Prepares Minden Biotech for migration of database programs
- Expands the current Customer service capabilities
- Reduces expenses by increasing existing staff efficiencies

Our proposal includes:

- Integrating the existing PBXs, recommended Octel voice mail, and future database programs
- Considers Minden's concerns to limit transmission costs
- Combines multiple Interactive Voice Response, Fax Back, and Middleware functions into one Lynx function

The architecture of the Minden Biotech solution at the head office in Minden, Nevada is depicted in Figure 6-3 on page 272. Likewise, the New Jersey office call center diagram is depicted in Figure 6-4 on page 273 while the London, UK office is shown in Figure 6-5 on page 274. Nonetheless, Capri Systems has also provided the solutions as requested by the RFP, a discussion of the benefits and disadvantages of the solutions are discussed under the General discussion on the RFP response on page 316.

Capri Systems has an alternative solution, shown in Figure 6-6 on page 275, is a divergence to the request for proposal plan. This alternative shows the benefits of merging the Nevada, and New Jersey call centers into a single call center in Minden, Nevada. This solution is discussed further as well at the end of the chapter in a general discussion of other potential solutions.

---

[64] Portions copyright 1997 Capri Systems Inc., used with special permission.

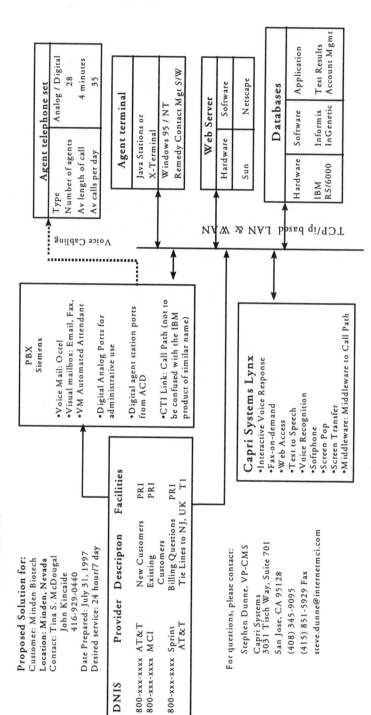

Figure 6-3 Minden Biotech Head Office proposed call center diagram

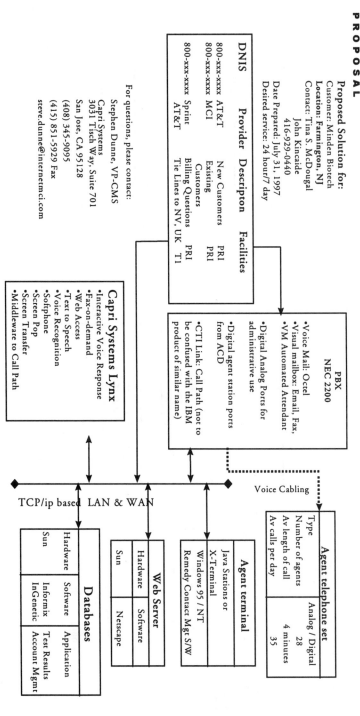

CAPRI SYSTEMS INC. RESPONSE TO THE MINDEN BIOTECH REQUEST FOR PROPOSAL

Proposed Solution for:
Customer: Minden Biotech
Location: Farmington, NJ
Contact: Tina S. McDougal
    John Kincaide
    416-929-0440
Date Prepared: July 31, 1997
Desired service: 24 hour/7 day

DNIS

DNIS	Provider	Description	Facilities
800-xxx-xxxx	AT&T	New Customers	PRI
800-xxx-xxxx	MCI	Existing Customers	PRI
800-xxx-xxxx	Sprint AT&T	Billing Questions Tie Lines to NV, UK	PRI T1

For questions, please contact:
Stephen Dunne, VP-CMS
Capri Systems
3031 Tisch Way, Suite 701
San Jose, CA 95128
(408) 345-9095
(415) 851-5929 Fax
steve.dunne@internetmci.com

PBX
NEC 2200

- Voice Mail: Octel
- Visual mailbox: Email, Fax, VM Automated Attendant
- Digital Analog Ports for administrative use
- Digital agent station ports from ACD
- CTI Link: Call Path (not to be confused with the IBM product of similar name)

Capri Systems Lynx

- Interactive Voice Response
- Fax-on-demand
- Web Access
- Text to Speech
- Voice Recognition
- Softphone
- Screen Pop
- Screen Transfer
- Middleware to Call Path

Agent telephone set

Type	Analog / Digital
Number of agents	28
Av length of call	4 minutes
Av calls per day	35

Voice Cabling

TCP/ip based LAN & WAN

Agent terminal

| Hardware | Java Stations or X-Terminal |
| Software | Windows 95 / NT Remedy Contact Mgt S/W |

Web Server

| Hardware | Sun |
| Software | Netscape |

Databases

Hardware	Software	Application
Sun	Informix InGenetic	Test Results Account Mgmt

Figure 6-4 Minden Biotech - New Jersey Call Center proposed call center diagram

273

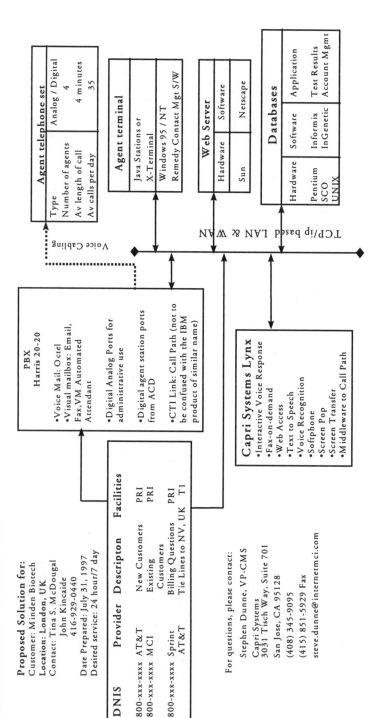

Proposed Solution for:
Customer: Minden Biotech
Location: London, UK
Contact: Tina S. McDougal
John Kincaide
416-929-0440
Date Prepared: July 31, 1997
Desired service: 24 hour/7 day

DNIS	Provider	Descripton	Facilities
800-xxx-xxxx	AT&T	New Customers	PRI
800-xxx-xxxx	MCI	Existing Customers	PRI
800-xxx-xxxx	Sprint	Billing Questions	PRI
	AT&T	Tie Lines to NV, UK	T1

For questions, please contact:

Stephen Dunne, VP-CMS

Capri Systems
3031 Tisch Way, Suite 701
San Jose, CA 95128
(408) 345-9095
(415) 851-5929 Fax
steve.dunne@internetmci.com

PBX
Harris 20-20

• Voice Mail: Octel
• Visual mailbox: Email, Fax,VM Automated Attendant

• Digital Analog Ports for administrative use

• Digital agent station ports from ACD

• CTI Link: Call Path (not to be confused with the IBM product of similar name)

Capri Systems Lynx
• Interactive Voice Response
• Fax-on-demand
• Web Access
• Text to Speech
• Voice Recognition
• Softphone
• Screen Pop
• Screen Transfer
• Middleware to Call Path

Voice Cabling

TCP/ip based LAN & WAN

Agent telephone set

Type	Analog / Digital
Number of agents	4
Av length of call	4 minutes
Av calls per day	35

Agent terminal

Java Stations or
X-Terminal
Windows 95 / NT
Remedy Contact Mgr S/W

Web Server

Hardware	Software
Sun	Netscape

Databases

Hardware	Software	Application
Pentium SCO UNIX	Informix InGenetic	Test Results Account Mgmt

Figure 6-5 Minden Biotech - London, UK proposed call center diagram

CAPRI SYSTEMS INC. RESPONSE TO THE MINDEN BIOTECH REQUEST FOR PROPOSAL

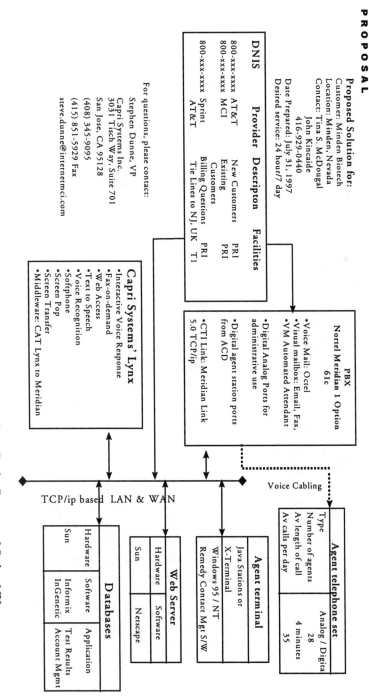

Proposed Solution for:
Customer: Minden Biotech
Location: Minden, Nevada
Contact: Tina S. McDougal
    John Kincaide
    416-929-0440
Date Prepared: July 31, 1997
Desired service: 24 hour/7 day

DNIS	Provider	Description	Facilities
800-xxx-xxxx	AT&T	New Customers	PRI
800-xxx-xxxx	MCI	Existing Customers	PRI
800-xxx-xxxx	Sprint	Billing Questions	PRI
	AT&T	Tie Lines to NJ, UK	T1

For questions, please contact:
Stephen Dunne, VP
Capri Systems Inc.
3031 Tisch Way, Suite 701
San Jose, CA 95128
(408) 345-9095
(415) 851-5929 Fax
steve.dunne@internetmci.com

PBX
Nortel Meridian 1 Option
61c

•Voice Mail: Octel
•Visual mailbox: Email, Fax,
•VM Automated Attendant

•Digital Analog Ports for administrative use

•Digital agent station ports from ACD

•CTI Link: Meridian Link
5.0 TCP/ip

Capri Systems' Lynx
•Interactive Voice Response
•Fax-on-demand
•Web Access
•Text to Speech
•Voice Recognition
•Softphone
•Screen Pop
•Screen Transfer
•Middleware: CAT Lynx to Meridian

Voice Cabling

TCP/ip based LAN & WAN

Agent telephone set

Type	Analog / Digital
Number of agents	28
Av length of call	4 minutes
Av calls per day	35

Agent terminal
Java Stations or
X-Terminal
Windows 95 / NT
Remedy Contact Mgt S/W

Web Server

Hardware	Software
Sun	Netscape

Databases

Hardware	Software	Application
Sun	Informix	Test Results
	InGenetic	Account Mgmt

Figure 6-6 Minden Biotech: Proposed Alternative for USA - Centralized call center at Minden, NV

OFFICE CALL PROCESSING

Capri Systems has proposed an independent third party voice messaging system from Octel. This system will interface with the corporations existing multi-vendor PBX equipment environment. This system can be used for the following functions:

- Enterprise voice messaging with a full featured messaging environment that meets the criteria set out in the RFP.

- Degree of integration is dependent upon the current software releases of the PBXs and their forms of signaling (i.e. Inband DTMF, or serial or TCP/ip based interfaces).

- Message waiting support for those phones that can be supported.

Octel Communications has surpassed all others in two areas:

1. Octel provides referenceable PBX interfaces to the Siemens, NEC, and Harris PBX's

2. Octel provides an effective graphical interface to display faxes, e-mails, and voice messages in a Windows environment

As a result, Capri Systems suggests that the fax messaging application be separated from the fax back applications associated with the lab results requirements of this RFP.

CALL CENTER INTEGRATION SOFTWARE FOR INGENETIC APPLICATION

Capri Systems has our own middleware running on the Sun server for the Nortel Meridian Link CTI. For the other systems including Lucent, we recommend Nabnasset. These products would be used to help integrate the call handling and screen pop spawning of the Informix based applications.

Working with major utilities, financial institutions, and high technology businesses, Capri Systems has discovered that most call centers have less than 50 agents.

Recent nationwide research performed by CAT Technology of San Jose, California surveyed over 300 businesses who provided the following observations:

- 54% of call centers have less than 50 agents
- 46% handle less than 10,000 calls per month
- 40% of call center administrators plan to spend as much as $350,000 on improvements in the next 18 months

Overall, the major concern of call center managers is how to cost-effectively integrate existing databases, PBX's, and agent desktop devices.

Capri Systems has met this challenge with Lynx 35/75/1000. Lynx offers all the major call center features, including PBX integration middleware, in a single cost-efficient Sun workstation. These features include:

- Interactive voice response
- Voice recognition
- Database repository
- Web server
- Fax server
- Text-to-speech
- Soft phone
- Screen pop

Features and Functions

Lynx offers

- 24 hour, 7 day a week phone answering and management
- Integrates to existing PBX and host database management system
- Custom applications
- Real time and historical reporting

- Client/server architecture
- Ethernet TCP/IP networking

Telephony Functions

- Log on/off
- Hold
- Conference
- Supervisor monitoring
- Caller profiles
- Voice response system activity
- Supervisor Broadcast
- Call transfer
- Event notification
- Multiple concurrent phone calls
- Skill based routing

JAVA Applications

- Personal computer, network computer, or JAVA station for Customer service agents
- Personal computer, network computer, or JAVA station for supervisors
- Graphical user interface (GUI) applications for Customer service rep stations
- Home agent and telecommuter support via standard browser technology

Benefits

- Scalability
- Reliability
- Ease of use
- Ease of installation
- Ease of customization for specific Customer applications
- Higher Customer satisfaction

- Increased employee satisfaction
- Stronger database integrity
- Added agent productivity
- Added preparedness
- Better business
- Lower costs
- Optimized computing and telephony infrastructure

Lynx 35/75/1000

Capri Systems has pre-packaged configurations depending on:

Number of agents

Type of PBX and ACD

Complexity of applications

Growth expectations

Features and Capabilities

- caller by Automatic Number Identification, Dialed Number Identification Service, or Customer-supplied identification number
- Agent scripting information
- Call and caller information
- Directs callers through menu selections
- Answers questions automatically or transfers to live Customer Representative
- Looks up information from the database and reports it to the caller and / or agent
- Automated routing
- Registers data and maintains a file of number of calls, duration, length of time in queue, etc.
- Captures voice messages from callers in conjunction with PBX voice mail
- Automatically returns FAX messages to the caller where requested

- Connects to applications servers
- Enhanced call wrap-up
- Call traffic and resource utilization
- Activity tracking by individual or Customer group
- Browse reports of Customer profile and annotations
- Details product-related calls and activities

Equipment

Lynx is designed to operate on a Sun Microsystems workstation which may be provided by Capri Systems or the Customer including:

- Solaris 2.X UNIX operating system
- Sun Microsystems Ultra Server hardware platform
- PC running Windows 3.x and Windows NT
- PC running SCO UNIX

Interactive voice response, fax, text-to-speech, and speech recognition functions are accomplished via Linkon S-Bus expansion boards.

PBX interfaces will be provided in accordance with your PBX vendor. Capri Systems will coordinate installation.

Database repository may be your legacy systems or Capri Systems will provide as part of our integration, depending on your environment.

Project management, custom software design, and scope definitions are handled by Capri Systems' experienced engineering and program management personnel.

An evaluation of a successful implementation will include:

- Reliable statistics
- Measurable performance

- Better forecasting

- Full equipment utilization

- Optimized employee performance

INTERNET/INTRANET LONG DISTANCE SERVICES OR LONG DISTANCE COST REDUCTION STRATEGY.

Capri Systems does not recommend at this time that Internet/Intranet based Internet protocol data transmission of voice as a strategy at this time. However, the costs of the voice lines is typically a lower percentage ongoing cost of a call center. Capri Systems recommends that Minden Biotech negotiate with their local and long distance carriers. Typically 7 to 8 cents per minute or even 6.5 cents minute is achievable with aggressive bargaining and commitment.

FAX SERVER

Capri Systems would recommend that the Lynx system provide the bulk and client based faxing services. Fax server technology for integrated messaging services would be best served using the Octel Communications fax mail services, which are independent of the Octel system.

INTEGRATION FOR ALL SITES

In order to describe some of the challenges posed by this RFP, one needs to review the components and the call flow.

For each call center, the calls would arrive and terminate in the PBX. A caller may elect, if they have the access to the service to use the fax-on-demand services, or if they want, branch to a live agent to discuss laboratory results. In each case the call is being managed by the Capri Systems Lynx system, which would either furnish the answers using the interactive voice response, or fax-on-demand services. The Lynx system would also interface with the PBX via middleware technologies from Nabnasset. Through this middleware technology, the Lynx system would be able to manage the communications between the agents and the database, in order for the screen pops to occur. This would involve the ACD functions that are available to the legacy PBX systems at each location.

If a call needs to be handed off to voice mail, or a fax needs to be delivered for a voice mail box owner, then the Octel system would manage that office messaging system. Corporate email would be also integrated, and for those that have access to integrated messaging services they would be able to access all these message formats either via phone, or through the use of a computer system running Octel's integrated messaging services on Windows based technology.

The management advantages is that that no matter which site has what kind of PBX, and version of UNIX operating system, it will allow Minden Biotech to have a common management and computer telephony integration at each location. Once this foundation is laid then the integration of the US based call center could take place.

Phase I: Implementation Plan

The Phase I Implementation Plan would consist of the following steps.

- System analysis
- Resource assessment
- Work flow analysis
- Scope
- Quotation
- Contract
- Development
- Coordinate vendors
- Schedule
- Training
- Testing
- Evaluation

- Future planning

Standard Process for Implementation
1. Concept of Operation
2. Functional Specifications
3. Design Specifications
4. Development and Testing
5. Vendor Coordination - Telco
6. Vendor Coordination - PBX
7. Vendor Coordination - DBMS
8. Pilot on Line - 1-25 Agents
9. Pilot on Line - 25-49 Agents
10. Pilot on Line - 50-75 Agents
11. Train Administrators including Preparation
12. Travel & Lodging Surcharge per traveling day
13. Travel & Lodging Surcharge per stationary out of town day
14. Full System on Line
15. Evaluation
16. Pure Speech Applications Development
17. Web Development
18. Future Planning

PROJECT MANAGEMENT

The software engineering team will be organized around a Program Manager, responsible for overall completion of the project on schedule and on budget. The program will be divided into three categories:

1. Applications Design Team
2. Development Team
3. Database Design, Data migration and Reports

Typical project phases include:

- Object Design
- Screen Design
- DB Design.
- Design Review
- Object/Method Coding and Testing
- Screen Coding and Testing
- Prototype Demonstration
- Data Migration
- Report Development
- Integration
- Test
- User Acceptance

Management Team

Paul Schaller, Chairman, who has over twenty years in the emerging computer and networking marketplace. Mr. Schaller was involved with the initial growth of Vitalink, Harmonic Lightwaves, Alantec, and Fore Systems.

Bob Smith, President, who has over twenty years of sales and marketing management experience with Sun Microsystems, Harris Corporation, Hewlett-Packard, and South Central Bell as well as general management experience in new ventures.

284

Stephen Dunne, Vice President of Sales and Marketing, who has twenty-one years experience in telecommunications including the last three years as one of the top senior account executive for Nortel in the nation.

John Morse, Vice President of Finance, who has over thirty years experience in the financial management of high technology companies.

Bob Williams, VP SE Sales, and Mike Ray, Director of SE Operations, manages the office in Norcross, Georgia.

Jon Pearce, PhD, Consultant and Professor of Mathematics and Computer Science, San Jose State University, holds a PhD in Mathematics from the University of California, Berkeley. He was Manager of Application Development Technology for the Silma Division of Adept Technologies, Inc. He has programming experience using UNIX, DOS, C, C++, Java, HTML, Sil, Pascal, MODULA2, Scheme, Miranda, Gofer, Prolog, and assembly language for the 80x86, MIPS, and PDP-11 processors.

PHASE II IMPLEMENTATION

Phase II of the implementation would go through much the same processes as described above, however, it would not necessarily take as long. The focus will be on the pilot project of the integration of both the voice and WAN resources. This is a key function that will need to be thoroughly tested and reviewed.

Capri Systems' Minden Biotech Project Pricing

The following pricing schedules **are manufacturers suggested retail pricing**, and reflect the needs of the Minden Biotech RFP for the original request for 3 integrated call centers. The pricing is current to late 2nd quarter 1997, when the RFP response was written. Capri Systems provided tables of other products such as the Recite speech recognition software that are not part of this RFP response. This is documented on page 287. They are listed to show what these products are and that they are available from them. Whenever you see a product that does not have a quantity value and resulting extended price, then those products are not being used in this RFP Response. The only exception to this situation is the table "Customer to Provide" on page 294. In that table, their is no pricing since these are quantities of products or items the customer is expected to provide for the facilitation of the installation of the Capri solution at each call center. You will find line item 108 "adjustments" on the last page of Exhibit "B". This is a customer discount to help win the deal. The part numbers for the each of the items are in a separate Table 6-3 Capri Systems - Part Number to Item Cross Reference - Minden Biotech RFP Response on pages 296-300.

Exhibit B: Lynx Hardware and Software

Quote Number:

Date: April 1, 1997

To: John Kincaide
Consultant
Minden Biotech
Minden, Nevada
Tel: 416-929-0440
Fax: 416-929-0059
FOB: San Jose, CA

Capri Systems LYNX CTI System
Sales Quotation to Attached Letter

From: Stephen Dunne
Vice President-CTI
Capri Systems, Inc.
3031 Tisch Way; Suite 701
San Jose, CA 95128
Tel: 408-345-9095
Fax: 415-851-5929

Expressed in U.S. Dollars. Terms: 25% upon signing, Net 30 plus 1.5% per month for past due amounts. Valid for 30 days from "Quote Date" above.

GENERAL PROJECT PURPOSE

Minden Biotech wishes to improve the responsiveness of three existing call center operations. Each location (Nevada, New Jersey, and London) have different PBX systems. As a result, Capri Systems is utilizing computer telephony integration middleware that standardizes the interface to the PBX link and TCP/IP network. Network diagrams have been included that show the relative connectivity among systems. Sun Microsystems will be utilized to run this middleware as well as the interactive voice response, fax back, and web access functions. Capri Systems recommends that Minden Biotech utilize the Octel voice mail platform because of the multiple voice mail interfaces available to the multiple PBX systems in use company wide.

Item	Description	Unit Price	Nevada	New Jersey	London	Total Qty	Ext Price
	Sun Microsystems						
1	Sparc 5 Workgroup Server, 17" Monitor	$ 7,995	1	1	1	3	$23,985
2	Ultra Enterprise 1, 20" Monitor	$ 19,995	1			0	$0
3	Ultra Enterprise 2, 20" Monitor	$ 29,995	1			0	$0
4	Internal Floppy Disk	$ 150	1	1	1	3	$450
5	North American Country Kit, Keyboard, Powercord, Mouse	$ -	1	1	1	3	$0
6	4-8 Gbyte 4mm DDS-2 Tape Unipack	$ 1,725	1	1	1	3	$5,175
7	CD-Rom Media for Solaris	$ 100	1	1	1	3	$300

Linkon Voice Processing Equipment

Item	Description	Voice	Voice/Fax	Voice/Fax/Speech	Telco	DSP's	S-Bus Slots	Incre-ments	RAM	Unit Price	Nevada	New Jersey	London	Total Qty	Ext Price
8	Integrix 6-slot system with connector to Sun						6			$ 1,995	1	1	1	3	$5,985
9	FS4000 driver for Sparc Solaris 2.5									$ 1,275	1	1	1	3	$3,825
10	T-1 interface w/out CSU (Sprite/Newbridge)				24		1			$ 5,500				0	$0
11	4 channel analog telco (Req's BBR, 03-AIC)		4		4	4	2		4	$ 3,095	2	2	1	5	$15,475
12	4 ch hicap analog board (Req's BBR, 03-AIC)			4	4	4	2		16	$ 5,950				0	$0
13	Breakout box for 2-wire loop start tel cable									$ 50	2	2	1	5	$250
14	Analog interface cable									$ 65	2	2	1	5	$325
15	Caller ID per analog telco port only							4		$ 100	8	8	4	20	$2,000
16	1 channel digital board		1		1	1	2		1	$ 1,393				0	$0
17	4 channel hi-performance digital board			4		4	2		16	$ 3,913				0	$0
18	8 channel hi-capacity digital board			8		8	2		32	$ 10,360				0	$0
19	8 channel hi-performance digital board		8			8	2		8	$ 7,126				0	$0
20	24 channel hi-density voice only digital board	24				8	2			$ 7,126				0	$0
21	FAX RTU per accessed VRU channel							4		$ 250	8	8	4	20	$5,000
22	Text to speech Am. English per VRU channel							4		$ 450	8	8	4	20	$9,000
23	TeraVox developer's license									$ 5,000				0	$0
24	TeraVox Run-Time License-all VRU channels									$ 150	8	8	4	20	$3,000
25	Voicetek - Generations (Per related VRU port)									$ 300				0	$0

Item	Description	Unit Price		Nevada	New Jersey	London	Total Qty	Ext Price
26	Firewall	$	4,990					$0
27	Oracle Enterprise Software	$	11,960					$0

Item	Description	Unit Price		Nevada	New Jersey	London	Total Qty	Ext Price
	Pure Speech Voice Recognition Products (Requires Sun Ultra 1 or 2)							
28	Recite - Digits RTU (1 RTU : 1 total VRU ports)	$	800				0	$0
29	Recite - Lite Version RTU (1 RTU : 1 total VRU ports) - Digits, 100 Words	$	1,800				0	$0
30	Recite - Pro Version RTU (1 RTU : 1 total VRU ports) - Digits, 2,000 Words	$	2,600				0	$0
31	Recite - IQ Version (Future) RTU (3 Ports per 1 RTU)						0	$0
31	Speech Server (May be same Sun server. Not req'd for Voicetek, Periphonics, Syntellect)	$	7,500				0	$0
32	Toolkit including 2 PSE20-1.2 RTU's	$	7,500				0	$0

Item	Description	Unit Price	Nevada	New Jersey	London	Total Qty	Ext Price
	Capri Systems Warranty and Maintenance Service						
33	CTI PBX Link Software - First 20 agents	$ 25,000	1	1	1	3	$75,000
34	CTI PBX Link Software - Expand 20 to 25 agents	$ 5,000				0	$0
35	CTI PBX Link Software - Each additional 10 agents over 25 and up to 95	$ 2,500				0	$0
36	CTI PBX Link Software - Each additional 20 agents over 95	$ 2,000				0	$0
37	CTI PBX Link Software - Interactive Voice Response driver	$ 9,000				0	$0
38	CTI PBX Link Software - Installation	$ 6,000				0	$0
39	CTI PBX Link Maintenance Support	$ 9,000	1	1	1	3	$27,000
40	Linkon Unlimited Telephone Support - Overnight Board Return	$ 4,900				0	$0
41	Linkon 20 Incidents per year Telephone Support - Overnight Board Return	$ 2,900	1	1	1	3	$8,700
42	Voicetek Basic Care: 8 hour / 5 day plus upissues	$ -				0	$0
43	Voicetek Total Care: Support 24 hour / 7 day plus upissues	$ -				0	$0
44	Voicetek Soft Care: 24 hour / 7 day including software releases	$ -				0	$0
45	Pure Speech Support 24 hour / 7 day - First Year	$ 9,375				0	$0
46	Pure Speech Support 24 hour / 7 day - Second Year	$ 6,250				0	$0
47	Sun Telephone 8AM-8PM Onsite Hardware Support	$ 2,393				0	$0
48	Sun Telephone 24 Hour / 7 Day / 2 Hour 8AM-8PM Onsite Hardware Support	$ 3,733				0	$0
49	First year Silver Level Oracle Support	$ 2,880				0	$0
50	Shipping and Order Coordination	$ 500	1	1	1	3	$1,500
51	First Year Support	$ 8,501				1	$8,501

Item	Description	Unit Price	Nevada	New Jersey	London	Total Qty	Ext Price
	MetaSound Systems Recorded Announcement Device						
52	Digital Player with handset, voice module with one recordable RAN, licensed music	$ 627	1	1	1	3	$1,881
53	Voice Module - Not user recordable	$ 49				0	$0
54	Voice Module - User recordable	$ 79				0	$0
55	Voice Module - User recordable, Expanded Memory	$ 99				0	$0
55	Licensed Music Module	$ 49				0	$0
55	RCA-type External Music Module for Customer-provided music source	$ 199				0	$0
56	Customized professional voiceover per Customer script	$ 375				0	$0
57	Digital Studio	$ 699				0	$0
58	Module Programmer and Utility Software	$ 1,999				0	$0
59	Module Duplicator	$ 899				0	$0
60	Installation and Support	$ 250	1	1	1	3	$750
61	Annual Maintenance	$ 282	1	1	1	3	$846

INTEGRATED SOLUTIONS

Item	Description	Date	Days	Hourly Rate	Nevada	New Jersey	London	Total Qty	Ext Price
	Capri Systems Engineering and Project Management			Total Engineering Hours:				319.0	
62	Concept of Operation	9/11/97	3.8	$ 165	10	10	10	30	$4,950
63	Functional Specifications	9/14/97	1.9	$ 165	5	5	5	15	$2,475
64	Design Specifications	9/19/97	3.8	$ 165	10	10	10	30	$4,950
65	Development and Testing	9/23/97	1.9	$ 165	5	5	5	15	$2,475
66	Vendor Coordination - Telco	9/25/97	1.9	$ 165	5	5	5	15	$2,475
67	Vendor Coordination - PBX	9/30/97	1.9	$ 165	5	5	5	15	$2,475
68	Vendor Coordination - DBMS	10/02/97	3.8	$ 165	10	10	10	30	$4,950
69	Pilot on Line - 1-25 Agents	10/06/97	3.8	$ 165	10	10	10	30	$4,950
70	Pilot on Line - 25-49 Agents	10/09/97	0.0	$ 165				0	$0
71	Pilot on Line - 50-75 Agents	10/12/97	0.0	$ 165				0	$0
72	Train Administrators including Preparation	10/12/97	1.9	$ 165	5	5	5	15	$2,475
73	Travel & Lodging Surcharge per travelling day	10/14/97	0.8	$ 500	2	2	2	6	$3,000
74	Travel & Lodging Surcharge per stationary out of town day	10/15/97	1.9	$ 300	5	5	5	15	$4,500
74	Full System on Line	10/17/97	3.8	$ 165	10	10	10	30	$4,950
75	Evaluation	10/29/97	1.9	$ 165	5	5	5	15	$2,475
76	Pure Speech Applications Development	10/31/97	0.0	$ 185				0	$0
77	Web Development	10/31/97	5.6	$ 165	15	15	15	45	$7,425
78	Future Planning	11/15/97	1.3	$ 165	10			10	$1,650

Item	Description	Hourly Rate	Hrs	Ext Price
	Documentation			
79	Documentation including all the items listed below:	$ 165	3	$495
80	General design description written out in readable form	1	0	
81	General graphical description of the proposed Customer network		0	
82	Administrator guide for adding agents, adding fields, adding records, etc.		0	
83	Instructions for agents on how to access the system	1	0	
84	Complete software and hardware inventory list		0	
85	Complete documentation as provided by Sun Microsystems		0	
86	Contact lists for warranty assistance		0	
87	Complete documentation on extracting reports		0	
88	Complete warranty and maintenance coverages and procedures description		0	
89	Instructions on Internet Service Provider (ISP) access and services		0	
90	Instructions on common equipment to be ordered for Centrex or PBX ACD		0	
91	Instructions on PC hardware and browser requirements for agents		0	

Item	Customer Requirements	Nevada	New Jersey	London	Total Qty
	Customer to provide:				
92	Power requirements: 20 amp, 120VAC, isolated ground	1	1	1	3
93	Secure storage area for shipped hardware and software	1	1	1	3
94	Analog PBX station ports to customer provided PBX to access Capri hardware	1	1	1	3
95	Trunk side T-1 digital trunk interface to customer provided PBX to access Capri hardware				0
96	Line side T-1 digital trunk interface to customer provided PBX to access Capri hardware				0
97	CSU for T-1 digital trunk interface to customer provided PBX to access Capri hardware				0
98	Telephone set at hardware installation location to allow direct inward and outgoing calls during installation	1	1	1	3
99	Sufficient countertop area for installation of Capri hardware	1	1	1	3
100	Sufficient air circulation and environmental conditions to support Capri hardware	1	1	1	3
101	Measured business line or direct inward dial analog station port to remotely access modem connected to Capri hardware	1	1	1	3
102	Reasonable access to customer personnel to complete concept of design and schedule installations	1	1	1	3
103	Access to customer local area network and assigned IP addresses where appropriate	1	1	1	3
104	Letter of agency to allow contact with PBX, Centrex, and telco provider for placing needed orders or coordinate installation	1	1	1	3
105	Authorization for supervised or un-supervised access to customer premises for on site installation access	1	1	1	3
106	Contact lists for customer employees directly involved with this project	1	1	1	3
107	All associated inside building cabling including local area network and telco access	1	1	1	3

CAPRI SYSTEMS INC. RESPONSE TO THE MINDEN BIOTECH REQUEST FOR PROPOSAL

Item	Description	Qty	Adj.
	Adjustments		
108	Adjustments	1	-$49,319

Total excluding sales tax and shipping for all items unless otherwise specified: $206,299

Customer Acceptance upon signature:

x
Signature

x
Printed Name   Purchase Order Number:

x
Printed Title

(   )   -
Telephone Number   Ext: x

Please fax to 415-851-5929 upon signing.

Capri Systems Authorization:

_____ Date:

_____ Date:

Questions may be directed to:
Stephen Dunne (408) 345-9095

Capri Systems reserves the right to review all Sales Quotations that are dramatically altered as a result of the Concept of Operation, Functional Specifications, and Design Specifications phases of this project.

Table 6-2 Capri Systems' RFP Response: Exhibit B: Lynx Hardware & Software

Part number to Item Cross Reference Table

Item	Description	Part Number
	Sun Microsystems	
1	Sparc 5 Workgroup Server, 17" Monitor	S5FX1-110-32-P16
2	Ultra Enterprise 1, 20" Monitor	A11-UBA1-1A-064AB
3	Ultra Enterprise 2, 20" Monitor	A14-UCB1-1E-064AB
4	Internal Floppy Disk	X6001A
5	North American Country Kit, Keyboard, Powercord, Mouse	X3540A
6	4-8 Gbyte 4mm DDS-2 Tape Unipack	X6254A
7	CD-Rom Media for Solaris	SOLS-C
8	Integrix 6-slot system with connector to Sun	03-IN06
9	FS4000 driver for Sparc Solaris 2.5	04-LV0001-23
10	T-1 interface w/out CSU (Sprite/Newbridge)	03-NM2121-03
11	4 channel analog telco (Req's BBR, 03-AIC)	03-FS4004-A2
12	4 ch hicap analog board (Req's BBR, 03-AIC)	03-FS4004-A3
13	Breakout box for 2-wire loop start tel cable	BBR
14	Analog interface cable	03-AIC
15	Caller ID per analog telco port only	04-LV0006-00
16	1 channel digital board	04-LV0009-09
17	4 channel hi-performance digital board	03-FS4004-D2
18	8 channel hi-capacity digital board	03-FS4008-D3
19	8 channel hi-performance digital board	03-FS4008-D2
20	24 channel hi-density voice only digital board	03-FS4024-D1
21	FAX RTU per accessed VRU channel	04-LV0009-09
22	Text to speech Am. English per VRU channel	04-LV0011-01
23	TeraVox developer's license	04-LV0001-03
24	TeraVox Run-Time License-all VRU channels	04-LV0002-03
25	Voicetek - Generations (Per related VRU port)	Voicetek-RSP/TSP
26	Firewall	Sun-Firewall
27	Oracle Enterprise Software	Oracle-DB
	Pure Speech Voice Recognition Products (Requires Sun Ultra 1 or 2)	
28	Recite - Digits RTU (1 RTU : 1 total VRU ports)	PSE00-1.2

Part number to Item Cross Reference Table

Item	Description	Part Number
29	Recite - Lite Version RTU (1 RTU : 1 total VRU ports) - Digits, 100 Words	PSE10-1.2
30	Recite - Pro Version RTU (1 RTU : 1 total VRU ports) - Digits, 2,000 Words	PSE20-1.2
31	Recite - IQ Version (Future) RTU (3 Ports per 1 RTU)	TBD
31	Speech Server (May be same Sun server. Not req'd for Voicetek, Periphonics, Syntellect)	PSK100-1.2
32	Toolkit including 2 PSE20-1.2 RTU's	PSK100-1.2
	Capri Systems Warranty and Maintenance Service	
33	CTI PBX Link Software - First 20 agents	CAT-CTI-1-20
34	CTI PBX Link Software - Expand 20 to 25 agents	CAT-CTI-Add5/21-25
35	CTI PBX Link Software - Each additional 10 agents over 25 and up to 95	CAT-CTI-Add10/26-95
36	CTI PBX Link Software - Each additional 20 agents over 95	CAT-CTI-Add20/95+
37	CTI PBX Link Software - Interactive Voice Response driver	CAT-CTI-IVR-DRVR
38	CTI PBX Link Software - Installation	CAT-CTI-Install
39	CTI PBX Link Maintenance Support	CAT-CTI-Mtce
40	Linkon Unlimited Telephone Support - Overnight Board Return	08-Premier-YR
41	Linkon 20 Incidents per year Telephone Support - Overnight Board Return	08-Enhanced-YR
42	Voicetek Basic Care: 8 hour / 5 day plus upissues	Voicetek-Spt-Basic
43	Voicetek Total Care: Support 24 hour / 7 day plus upissues	Voicetek-Spt-Total
44	Voicetek Soft Care: 24 hour / 7 day including software releases	Voicetek-Spt-Soft
45	Pure Speech Support 24 hour / 7 day - First Year	PS-Spt-Yr-1
46	Pure Speech Support 24 hour / 7 day - Second Year	PS-Spt-Yr-2
47	Sun Telephone 8AM-8PM Onsite Hardware Support	Spectrum-Silver
48	Sun Telephone 24 Hour / 7 Day / 2 Hour 8AM-8PM Onsite Hardware Support	Spectrum-Gold
49	First year Silver Level Oracle Support	PSE20-1.2

Part number to Item Cross Reference Table

Item	Description	Part Number
50	Shipping and Order Coordination	CAT-SHP-1
51	First Year Support	CAT-SUP-1
	MetaSound Systems Recorded Announcement Device	
52	Digital Player with handset, voice module with one recordable RAN, licensed music	DMOH-1000
53	Voice Module - Not user recordable	VM-8.0
54	Voice Module - User recordable	M-8.1
55	Voice Module - User recordable, Expanded Memory	VM-16.1
55	Licensed Music Module	MM-1
55	RCA-type External Music Module for Customer-provided music source	EMM-1
56	Customized professional voiceover per Customer script	CAT-MOH-VO-1
57	Digital Studio	DDOH-5000
58	Module Programmer and Utility Software	MMP-1000
59	Module Duplicator	MD-1000
60	Installation and Support	CAT-MOH-Inst-1
61	Annual Maintenance	CAT-MOH-Mtce-1
	Capri Systems Engineering and Project Management	
62	Concept of Operation	CAT-Eng 1
63	Functional Specifications	CAT-Eng 2
64	Design Specifications	CAT-Eng 3
65	Development and Testing	CAT-Eng 4
66	Vendor Coordination - Telco	CAT-Coord-Tel
67	Vendor Coordination - PBX	CAT-Coord-PBX
68	Vendor Coordination - DBMS	CAT-Coord-DB
69	Pilot on Line - 1-25 Agents	CAT-Eng 5
70	Pilot on Line - 25-49 Agents	CAT-Eng 6
71	Pilot on Line - 50-75 Agents	CAT-Eng 7
72	Train Administrators including Preparation	CAT-Eng 8
73	Travel & Lodging Surcharge per traveling day	CAT-TL-Tvl
74	Travel & Lodging Surcharge per stationary out of town day	CAT-TL-Sta
74	Full System on Line	CAT-Eng 9
75	Evaluation	CAT-Eng 10

Part number to Item Cross Reference Table

Item	Description	Part Number
76	Pure Speech Applications Development	CAT-PS-Dev
77	Web Development	CAT-WEB-35
78	Future Planning	CAT-Eng 11
	Documentation	
79	Documentation including all the items listed below:	CAT-Doc-Kit
80	General design description written out in readable form	CAT-Doc 1
81	General graphical description of the proposed Customer network	CAT-Doc 2
82	Administrator guide for adding agents, adding fields, adding records, etc.	CAT-Doc 3
83	Instructions for agents on how to access the system	CAT-Doc 4
84	Complete software and hardware inventory list	CAT-Doc 5
85	Complete documentation as provided by Sun Microsystems	CAT-Doc 6
86	Contact lists for warranty assistance	CAT-Doc 8
87	Complete documentation on extracting reports	CAT-Doc 9
88	Complete warranty and maintenance coverages and procedures description	CAT-Doc 10
89	Instructions on Internet Service Provider (ISP) access and services	CAT-Doc 11
90	Instructions on common equipment to be ordered for Centrex or PBX ACD	CAT-Doc 12
91	Instructions on PC hardware and browser requirements for agents	CAT-Doc 13
	Customer to Provide	
92	Power requirements: 20 amp, 120VAC, isolated ground	
93	Secure storage area for shipped hardware and software	
94	Analog PBX station ports to customer provided PBX to access Capri hardware	
95	Trunk side T-1 digital trunk interface to customer provided PBX to access Capri hardware	
96	Line side T-1 digital trunk interface to customer provided PBX to access Capri hardware	

Part number to Item Cross Reference Table

Item	Description	Part Number
97	CSU for T-1 digital trunk interface to customer provided PBX to access Capri hardware	
98	Telephone set at hardware installation location to allow direct inward and outgoing calls during installation	
99	Sufficient countertop area for installation of Capri hardware	
100	Sufficient air circulation and environmental conditions to support Capri hardware	
101	Measured business line or direct inward dial analog station port to remotely access modem connected to Capri hardware	
102	Reasonable access to customer personnel to complete concept of design and schedule installations	
103	Access to customer local area network and assigned IP addresses where appropriate	

Table 6-3 Capri Systems - Part Number to Item Cross Reference - Minden Biotech RFP Response

Capri Systems History and Experience

Capri Systems is located in San Jose, California, and has recently opened its Atlanta office in Norcross, Georgia.

Capri Systems, a Nortel Business Affiliate, specializes in systems integration and application development for the Customer Management Solutions (CMS) and Computer Telephony Integration (CTI) market using the following technologies:

- Sun SPARC/Solaris Client Server/Network Computing

- PC Platforms including Windows NT/3.x and SCO UNIX

- Internet/Intranet/Java Programming Language/Java Network Computer

- Relational Database Management System - Oracle/Informix/Ingres

- Call Center/Help Desk Contact Management Software

Capri Systems sells systems and develops custom applications for Customer management through the integration of computer, database and telecommunications systems. The solutions utilize computer-telephony integration (CTI), Relational Data Base Management Systems (RDBMS), JAVA, Internet, and Intranet technology to create customized software applications for Customer management solutions. CAPRI SYSTEMS's integrated systems provide a wide variety of functions including:

- Client-Server computing
- Legacy RDBMS integration
- Java Internet/Intranet design
- PBX to computer interfaces
- PBX to central office interfaces
- Automated screen pop
- Skill based routing
- Interactive voice response
- Automated FAX
- Text-to-Speech
- Voice Recognition

Capri Systems maintains partnerships with vendors supplying the following products:

- Sun Microsystems
- Nortel Meridian 1 PBX systems
- Nabnasset call center middleware
- Voicetek Interactive voice response software
- Linkon voice response hardware and software
- Oracle
- Informix

- Remedy

Integrated customer management systems are complex products, requiring the coordination of numerous complicated and highly technical hardware and software products. In addition, customers usually require substantial consulting, product customization, and support.

Capri Systems Experience

Capri Systems provides a team of software engineers and consultants who average over fifteen years of experience in complex information technology environments. Our team is highly experienced in projects requiring the following:

- General Systems Integration & Development
- Custom C, C++ programming
- Custom JAVA programming
- Hardware installation
- Consulting services
- System management
- Installation of disk mirroring software
- Installation of Web/RDBMS monitoring software
- SQL database access
- Web based client front ends to systems utilizing interactive voice response.
- Back up and restore
- Recover and troubleshooting
- Analyze and tune for optimum performance
- File and login security procedures
- System startup and shutdown
- Adding new devices and software
- Installing new software revisions
- Systems upgrades
- Monitor network traffic and reconfigure for optimum performance
- Computer Telephone Integration Applications & Development

- Computer Telephone Integration Analysis and Design
- Customized programming of CTI Server to PBX
- Interactive voice response system installation, programming, and management
- ACD customization, design, and installation
- Call Center Software installation and management
- Programming link between CTI Server and DBMS
- Prime contractor for turnkey CTI installations
- Installation and management of Interactive fax-back system
- Outbound dialer customization and support
- Applications server support
- Image processing installation and support
- Auto attendant and other voice processing design and support
- Link of CTI system to mainframe
- GUI Programming
- Object Oriented Design and Programming

Sample Capri Systems Purchase and License Agreement

The following *sample* contract is provided as an example of the terms and conditions used in the industry. A word of warning! This is an example and not a contract that should be used for business transactions. Consult your qualified legal advisor about terms and conditions of any agreement. This agreement shown below is not to be used in any transaction.

CAPRI   SYSTEMS   PURCHASE   AND   LICENSE AGREEMENT

This Purchase and License Agreement (this "Agreement") is entered into by and between Capri Systems, Inc. ("Seller"), and _____("Customer").

Capri and Customer hereby agree as follows:

1.   PURCHASE, PRICE AND PAYMENT TERMS

a)   Purchase. Seller agrees to sell and Customer agrees to buy the "System" consisting of the hardware ("Hardware"), and the licensed software ("Software") listed in Exhibit A attached hereto. Seller's obligations under this Agreement are subject to Seller's credit approval of Customer.

b)   Price. The purchase price and license fees for the System and the services described in this Agreement is the amount specified in Exhibit A as the "Purchase Price." The Purchase Price does not include applicable taxes. In addition to the Purchase Price Customer is responsible for the payment of all taxes applicable to this sale or Seller's performance of this Agreement, except for any tax on Seller's net income.

c)   Payment Terms The Purchase Price will be due and payable as described in Exhibit A. Any invoice not paid in full within 30 days of its date will be past due and Customer agrees to pay late payment charges equal to one and one-half percent (1-1/2%) of the unpaid balance per month (or the maximum rate allowed by the law, if lower). Neither the invoicing of late payment charges nor the acceptance of them is to be deemed an agreement to either extend or finance payments due or past due. In addition, Customer agrees to pay all collection costs and reasonable legal fees incurred by Seller as a result of Customer's late payment or non-payment.

2.   DELIVERY AND INSTALLATION

a)   Installation. The System will be installed by Seller at the "Installation Site" on the Installation Date set forth in Exhibit A. Customer shall allow Seller's employees, representatives and subcontractors reasonable access to the necessary premises for installation. Before and during installation Customer is responsible to ensure the timely and adequate delivery, installation and functioning of any electrical and telecommunications connections.

b)   Customer Caused Delay. If Customer causes a delay in the delivery of the System, Customer shall be responsible for storage and other costs incurred by Seller, and any installment of the Purchase Price due after the delay shall be due and payable on the date specified in Exhibit A. Additional charges may apply if Seller must perform extra services or bear additional costs (such as overtime wages) because of an unprepared installation Site or due to Customer's acts or omissions, or conditions at the Installation Site that Customer should have reasonably expected to impact Seller's ability to install the System, but which Customer failed to inform Seller about in writing prior to the execution of this Agreement by Seller.

3.   CHANGES IN SYSTEM CONFIGURATION

a) To make a change in an accepted order Customer must notify Seller in writing of the desired change. Any such request from Customer for a change to an order previously accepted by Capri may subject Customer to a price change reflecting the inclusion or substitution and/or Capri's cost of processing such change request. Such a change may also result in a modification to the delivery schedule set forth in Exhibit B. Changes are allowed which cumulatively do not increase the Purchase Price by more than 20% or decrease the Purchase price by more than 10%. Seller will advise Customer if the requested Change is not technically feasible or is otherwise outside the manufacturer's guidelines. If Seller accepts a requested change, Seller will send a written acceptance notice setting forth any modifications of the Purchase Price and/or the Project Schedule. Any increase or decrease in the Purchase Price shall be invoiced or credited as applicable. A restocking fee equal to 20% of the list price of any returned component that has been delivered to the Installation Site will also be invoiced as an additional charge.

4. TRAINING

a) Seller shall provide Customer with its standard user training for the System at no additional charge. The standard user training for a given system type consists of instructional materials, and may include training sessions with an instructor. Other materials and training are available at an additional charge.

5. LIMITED WARRANTIES

a) Seller warrants that for twelve (12) months after the date of Cutover ("Warranty Period"): (a) the Hardware shall be free from equipment defects and faulty workmanship; (b) the Software shall function substantially in accordance with the applicable published functional specifications; and (c) the installation of the System shall conform to the manufacturer's installation specifications (collectively referred to as the "Warranties"). Warranties related to any additions to the hardware of Software installed during the Warranty Period shall terminate at the end of the Warranty Period for the System.

b) The foregoing warranties are contingent upon Customer's proper use and service of the System in applications for which the System was intended. The foregoing warranties do not apply to: damage to the System due to abuse, misuse, neglect, unauthorized repair or installation, fire, explosion, power irregularities, power surges, Acts of God (including, without limitation, earthquakes, rains, floods or lightning), or use of software other than that supplied with the System; or any other cause not attributable to Seller.

c) Customer must notify Seller promptly of any claimed defect or failure of any of the Warranties. Seller's sole obligation and Customer's exclusive remedy for any defect or failure of any Warranty during the Warranty Period will be for Seller, at its discretion, to repair or replace any defective items. Such repair or replacement during the Warranty Period will not extend or restart the Warranty Period. If neither repair or replacement is commercially practicable, Capri Systems shall terminate this Agreement and refund to Customer monies paid hereunder.

d) THE LIMITED WARRANTIES DESCRIBED ABOVE IN THIS SECTION, AND THE REMEDIES FOR A FAILURE, DEFECT OR BREACH OF ANY OF THOSE LIMITED WARRANTIES WHICH ARE EXCLUSIVE AND PROVIDED TO CUSTOMER IN LIEU OF ALL OTHER WARRANTIES, WRITTEN OR ORAL STATUTORY, EXPRESS OR IMPLIED INCLUDING WITHOUT LIMITATION. THE IMPLIED WARRANTIES OF MERCHANTABILITY, FITNESS FOR A PARTICULAR PURPOSE, OWNERSHIP AND NONINFRINGEMENT, WHICH SELLER SPECIFICALLY DISCLAIMS.

6.   SOFTWARE LICENSE

a) Customer use the Software provided with or integrated into the Hardware under the applicable license terms established by the Software owner(s) ("Owner") and set forth in Exhibit B.

b) Where no Software License Addendum for particular Software is attached, but Software is provided with or is integrated into the applicable Hardware. Customer is granted a non-exclusive, nontransferable license to use such Software. In consideration of the license, Customer agrees that: (a) the Software is the Owner's property; (b) the Software will be used to operate the System, or applicable part of it, for its own internal business purposes only; (c) it will not reproduce the Software, except reverse engineer decompile disassemble or service source code from the Software; (e) it will return or destroy the Software, and all copies of the Software, once it is no longer needed or permitted for use with the System, or applicable part of it, and (f) it will treat the Software as confidential and not disclose the Software to third parties.

c) To the extent any Software is delivered to Customer packaged with a "shrink-wrap" license the terms of such shrink-wrap license shall govern Customer's rights to use such Software with the System or applicable part of it.

7.   INDEMNITIES

a) Each party shall indemnify the other with respect to any third party claim alleging bodily injury, including death, or damage to tangible property, to the extent such injury or damage is caused by the negligence or willful misconduct of the indemnifying party.

b) Seller shall also indemnify Customer with respect to any claim alleging that Customer's use of the System constitutes an infringement of any United States patent or copyright, if Seller has been notified and permitted to defend the suit as required by the following paragraph. If a court of competent jurisdiction issues an injunction against Customer prohibiting it from using the System because of such claim, Seller, at its option, shall either obtain for Customer the right to continue using the System, or replace or modify the System so that Customer's use is not subject to the injunction. If Seller cannot either acquire the right to use the System or replace or modify it in a commercially reasonable and timely manner, then Customer's remedy is to return the System to Seller (after giving written notice to Seller and receiving instructions for the return). If the system is returned neither party shall have any further obligation or liability under this Agreement, except that Seller shall refund the depreciated value of the system (excluding the value of the wiring and cabling portion thereof) as carried on Customer's books at the time of such return. This indemnity shall not apply to claims arising in respect to the use of the System in a manner not contemplated under this Agreement, or if the claims are based on the use of the System in conjunction with products not provided to Customer by Seller. This Section 7 sets forth Seller's entire obligation and Customer's sole remedy with respect to any infringement claims.

c) A condition precedent to any obligation of a party to indemnify shall be for the other party to promptly advise the indemnifying party of the claim and turn over its defense. The party being indemnified must cooperate in the defense or settlement of the claim, but the indemnifying party shall have sole control over the defense or settlement. If the defense is properly and timely tendered to the indemnifying party, then it must pay all litigation costs, reasonable attorney's fees, settlement payments and any damages awarded (but this may not be construed to require the indemnifying party to reimburse attorney's fees or related costs of the other party that the other party incurs either to fulfill its obligation to cooperate, or to monitor litigation being defended by the indemnifying party).is granted a license to

8.   RISK, TITLE, AND SECURITY AGREEMENT

a) Title to the Hardware shall pass to Customer when the Purchase Price has been paid in full. Risk of loss or damage to the System or any of its components shall pass to Customer upon delivery to the Installation Site. Until Customer pays the Purchase Price in full, Customer grants to Seller a purchase money security interest in the System and its proceeds. Seller's filing costs will be invoiced as an additional charge to Customer and Customer agrees to sign any financing statement or other document Seller considers necessary to protect Seller's rights under this security interest.

9.   CUSTOMER'S OBLIGATIONS

a) In addition to the obligations described in this Agreement, Customer shall timely complete the tasks identified as its duties in Exhibit B.

b) The Purchase Price is based in part upon the understanding that (a) Seller may use its own employees or subcontractor(s) of its choosing to perform all or some of its services; and (b) those area at the Installation Site where Seller's employees or subcontractors are required to work do not contain any asbestos or other hazardous material. If Seller is restricted by Customer in managing its utilization of employees or subcontractors, or if any asbestos or hazardous material exists at work sites, Seller may increase the Purchase Price to reflect increased costs and extend the time of performance to reflect reasonable additional time required to adjust for unanticipated activities. In addition, with respect to the presence of asbestos or other hazardous materials, Customer must, at its own expense, have the material removed or notify Seller to install the applicable portion of the system in areas at the Installation Site not containing such material. The Purchase Price does not include charges for doing installation work or performing other services outside Seller's normal work hours, except for a Cutover scheduled for an evening or weekend in Exhibit B. If Customer asks that certain work or services be done outside of Seller's normal work hours, or takes other actions that require such work, then Seller may increase the Purchase Price to reflect Seller's then current charges for work during such hours.

10.   DEFAULT AND REMEDIES

a) If any material breach of this Agreement continues uncorrected for more than 30 days after written notice from the aggrieved party describing the breach, the aggrieved party shall be entitled to declare a default and pursue any and all remedies available at law or equity, except as specifically limited elsewhere in this Agreement.   In addition, if Customer is the aggrieved party Customer may suspend its payment obligation relating to the breach until Seller's breach is corrected, and if Seller is the aggrieved party, Seller may suspend performance of its obligations until Customer's breach is corrected.

11.   LIMITATION ON LIABILITY

a) SELLERS ENTIRE LIABILITY UNDER THIS AGREEMENT, INCLUDING LIABILITY ARISING OUT OF THE SYSTEM PURCHASED, SERVICES PERFORMED OR FROM THE SELLER'S NEGLIGENT OR OTHER ACTS OR OMISSIONS, SHALL BE LIMITED TO THE PRICE OF THE SYSTEM AND/OR SERVICES GIVING RISE TO THE CLAIM REGARDLESS OF THE LEGAL OR EQUITABLE BASIS OF ANY CLAIM OR OF ACTUAL NOTICE, NEITHER SELLER NOR SELLER'S SUPPLIERS SHALL BE LIABLE FOR (A) ANY INCIDENTAL, INDIRECT, SPECIAL OR CONSEQUENTIAL LOSS OR DAMAGES OR (B) ANY

DAMAGES RELATING TO A CLAIM MADE AGAINST CUSTOMER BY A THIRD PARTY EXCEPT FOR INDEMNIFIED CLAIMS DESCRIBED IN SECTION 11. THESE LIMITATIONS SHALL REMAIN IN FULL FORCE AND EFFECT THROUGH ANY RENEWAL OF MAINTENANCE SERVICE PROVIDED FOR IN THIS AGREEMENT.

12. <u>GENERAL</u>

a) Customer warrants that the person signing this Agreement for Customer is authorized to do so, and that Customer has obtained all internal and external approvals and resolutions necessary to enter into this Agreement and make the Agreement binding upon Customer.

b) Neither party shall be liable for delays, loss, damages or other consequences of acts, omissions or events beyond a party's control and which may not be overcome by due diligence, or caused by strikes or labor strife and unrest.

c) This Agreement constitutes the entire agreement between the parties with respect to the described transaction. It supersedes all prior negotiations, proposals, commitments, advertisements, publications or understandings of any nature, whether oral or written. Any amendment or modification to this Agreement and any waiver of rights under this Agreement must be in writing clearly intending to modify or waive rights under this Agreement must be in writing representatives of both parties to be effective. In interpreting this Section it is agreed that any preprinted or added terms and conditions in a purchase order form or like forms used by Customer to implement or change System or product orders under this Agreement are void with respect to this Agreement, even if acknowledged in writing by Seller.

d) If any provision of this Agreement is held invalid, the remaining provisions shall continue in full force and effect and the parties shall substitute for the invalid provision a valid provision which most closely approximates the economic effect and intent of the invalid provision.

e) Customer may not assign or otherwise transfer this Agreement and its rights and obligations under this Agreement without the prior written consent of Seller, except that no such consent shall be required in the case of a merger or sale of all or substantially all of Customer's business assets related to this Agreement, provided that any assignee agrees in writing to be bound by the terms and conditions hereof. Any unauthorized assignment of this Agreement shall be null and void.

f) A failure by either party to exercise its rights under this Agreement shall not be a waiver of such rights.

g) To be effective any notification or consent required by this Agreement must be in writing and sent by prepaid certified mail, return receipt requested, to the addresses specified below or such other addressed as either party may specify by written notice. If the notice or consent is sent to Seller, then it should be marked to the attention of "Contracts Administration," and, if it is sent to Customer, then to the attention of the person signing this Agreement for Customer.

h) This Agreement shall be governed by the laws of the state of California without reference to conflict of laws provisions.

i) In the event Customer obtains financing for the purchase of the System from a finance leasing company acceptable to Seller, Customer may assign its right to receive title to the Hardware and delegate its payment obligations to such leasing company upon receipt of Seller's consent in the form provided by Seller. Even if Seller consents, it shall not be obligated to commence performance under this Agreement until it has received a written commitment from the leasing company to pay the Purchase Price under the terms set forth in this Agreement and has in fact made the payment of initial installment of the Purchase Price. When the Purchase Price (as adjusted) is paid in full, title to the Hardware shall vest in such leasing company.

j) This Agreement is not effective or binding upon Seller and does not constitute an offer subject to being accepted by Customer until it has been executed by a duly authorized representative of Seller. The effective date of this Agreement shall be the date of Seller's execution of this Agreement. Seller may deposit any check tendered by the Customer, but if Seller elects not to execute this Agreement, Seller shall promptly refund such amount to Customer. Any such deposit may not be construed as an acceptance or agreement by Seller to this Agreement becoming effective.

k) Exhibit A and Exhibit B are attached to and made part of this Agreement.

Accepted and agreed on   day of   ,   19
this

Customer:	CAPRI SYSTEMS, INC.
By:	By:
Name:	Name:
Title:	Title:

Exhibit A

1. Hardware: (See Exhibit B)

2. Software: (See Exhibit B)

3. Project Schedule: (See Exhibit B)

4. Scope of Work: (See Exhibit B)

3. Purchase
    Price:

$

Payment
Schedule
:

(a) 30% upon signing $

(b) 50 % upon System $
Delivery

(c) 20% upon Cutover $

4. Installation Date: (See Exhibit B)

5. Installation Site: (See Exhibit B)

Exhibit B

<u>Software License Terms and Conditions</u>

See Attachment labeled Exhibit B for Software Components.
Integrated Software Components may carry their own Software
License Agreements.
Documentation on those Software License Agreements is available upon
request.

General discussion on the RFP response

One of the interesting outcomes of working with Capri Systems on the RFP response was the review of the RFP with an alternative solution, which could reduce the size of the project initial cost, and increase the efficiencies or "critical mass" of the solution.

1. As Stephen Dunne said "bidders are sometimes a reticent in their willingness to suggest a complete different solution than one that was required in the RFP." The purpose of the RFP was more instructional than in one of pure economics. It strives to show how the integration of different call centers are pulled together into one cohesive organization. However, Stephen Dunne pointed out, it may be in Minden Biotech's best favor to cancel the idea of having a dedicated call center in New Jersey, and move the trained call center staff from there to the head office in Minden, Nevada. The production and some of the telephony functions (i.e. fax broadcasting of results) could stay local in New Jersey, but have all the voice inquires routed directly into the Nevada office: Since both sites will be on a WAN with production facilities at these locations, then to the outside world, the customer will still only see one company. Stephen suggested the following could happen:

2. The elimination of the Siemens switch and replace it with a more modern Nortel system. He recommends Nortel because of its features, and its general wider international support than do other switch manufacturers. Moreover, the Lynx system could use Capri Systems' own middleware solution which would provide much tighter integration with the Nortel and the InGenetic solution.

3. Scale of economics. With the smaller call centers in each location in the United States, then Minden Biotech would not get the scale of economy as it would with the centralization of the call center in Minden itself. By centralizing in one location, it is easier to manage, and Minden Biotech would reduce ongoing

telecommunication, and labor costs. Stephen estimates that the savings of the solution for the US, would a considerable amount of the initial cost.

4.   Logistics of call handling:

a)   Message waiting lights. In a multisite call center and operations environment, it is very difficult to initiate message waiting indication on phone sets between locations and between different PBX manufactures across the public switch telephone network. The best solution is to provide dedicated lines between locations, and this could be expensive.

b)   Screen pop transfers. This refers to an agent transferring a screen between call centers to another agent or supervisor. Internal screen pop transfers within a call center is an important feature, and is in demand more often. To do it between call centers is possible, yet again it would most likely require dedicated voice and data lines, that can be thoroughly tested and controlled. The major issue will be cost versus its need. If there is a need to do this only a few times each week, it may not be worth the extra money per month for communication lines and support of the systems.

Stephen's overall comment is that he positions UNIX based solutions because of reliability, and single platform management. His comments are that the customer really needs a stable and highly reliable platform, which UNIX provides. In each of the call centers in the RFP response, a single Sun workstation is managing all the major computer telephony functions. This eliminates the need for dedicated systems for interactive voice response that may be the case with other application platforms.

Other potential UNIX solutions for RFP

In this case, Capri Systems has used the Sun platform, and Linkon voice processing hardware. A different system integrator may have used an Intel based UNIX solution from which SCO, or Solaris may have been the focus. In either case, a system integrator so inclined would have produced a solution comparable to the Capri Systems' solution using Dialogic hardware and 3rd party technologies. Alternatively, Capri Systems could have provided the solution on a SCO OpenServer platform using a Pentium PC. Overall, the major challenge of the vendor solution is the ability to demonstrate it working at other sites. Moreover, there is an added factor that such companies will have to have the resources from which to try and test the integration of various products into a cohesive solution. This is a real challenge, and it should not be overlooked while reviewing solutions.

In the end, the purpose of the RFP was to raise the stakes quite high. I purposefully placed requested technologies such as PC switch, and Internet/Intranet telephony applications, which at the time of writing are just making themselves known and impacting the market. It's purpose here has been to educate rather than decide which is best or who has won. It is meant to provoke your thoughts on what is being offered, and moreover too show how these components work to develop a cohesive system. The lesson to be taken away is that because of nature of interconnecting two different networks together, you as a manager, should be aware that not one single system integrator, or even a single vendor may meet all your needs or goals. There sometimes have to be unavoidable compromises in terms of functionality, integration, and funding. The RFP's ideal is to have a single solution that would feed and integrate communications to a single point, where information would be platform independent, medium independent, and integrated on newer voice technologies. However, judicious review of your needs, and keeping an open mind that one of your vendors may suggest a completely new approach, (i.e. like centralizing all call center functions in the USA, rather than multi-site), is essential to success.

Case History Summary Tables

The following are the case history summary tables showing, cross-reference to case histories within the book, vendor companies, type of computer telephony application, vertical industry, and UNIX platform. All vendors have incredible and wonderful case histories to tell. Unfortunately, because of editorial constraints, they all could not be told. This is by no means to indicate that one case history was necessarily more compelling to tell than another. They all have a great value.

You will find **Vital Statistics Grand Summary Table** on page xx which summarizes the installation, cost factors, return on investment, ripple effect, and technologies deployed in the case histories.

A listing of all the vendor contacts are by company name in alphabetic order. This will allow you to contact the companies for further information.

I would also like to take the time to discuss how the case histories came to be selected, as well as the accuracy of the information within them.

The cases histories selected for write up were on a first come first served basis - more or less. We had to "trawl" the industry three times over a span from December 1995 to September 1996. We found that there was a large portion of sample had interactive voice response applications (40%), while the remaining were other computer telephony technologies. Since the sample is not scientifically randomly drawn then we can not really account for the large proportion of interactive voice response in the case history list. Based upon a hunch, my guess is that interactive voice response applications can be deployed with the minimal amount of computer to switch integration, making them easy to deploy. If this is true, then it is the author's sincere desire to see the success of the different computer telephony integration middleware APIs to succeed. This will break down some costs and barriers to the deployment of the technology.

The largest part of the case history writing effort started in earnest in late September, and October 1996. The technical writing and the fictitious Minden Biotech Request for Proposal was created from December 1996 to late January, 1997. A reply to this Request for proposal by Capri Systems started in late March 1997, and finished in mid April 1997.

Many different factors came to play as to which case histories were developed and how the RFP was created and answered. Time and the amount that the project demanded was the most influencing factor. We wanted to at least cover two case histories in each of the major computer telephony technology areas.

As to the accuracy of the information of the case histories, we have checked with the vendors and whenever possible the end users of the projects for each case history had been written. Most of the case histories have and both end user, and vendor, although this is not the case in every one. We feel there was no undue expression or influence brought to bear by the end users or vendors. The information herein is what has been reported to the author.

SO WE AGAIN APPLAUD THOSE WHO HAVE CONTRIBUTED!

The following table summarizes all the case histories by type of computer telephony application.

Case History Summary By Computer Telephony Application				
Case Page #	Company	Computer Telephony Application	Industry	UNIX
162	IBM USA	Call Center	Retail	AIX
255	Genesys	Call Center - Middleware	Financial	SUN
171	Teloquent	Call Center - Virtual	Utility	SCO
127	V-Systems	Fax Server	Aviation & Insurance	AIX, HP/UX, Sun
145	V-Systems	Fax Server	Financial	DG/UX
137	Faximum Software	Fax Sever	Manufacturing	SCO
99	CallStream	IVR	Transportation	SCO
117	MediaSoft	IVR	Banking	SCO
77	Voicetek	IVR	Help Desk Software	SCO & Sun
107	IBM Canada	IVR - Student Registration	Education	AIX
89	Parlance Corporation	IVR - Voice Recognition	Software	SGI IRIX
233	Arbinet	PC Switch	International CallBack	SCO
221	Cygnus	PC Switch	Marine Ship to Shore	QNX
241	AlphaNet	PC Switch	Intranet based Voice transport	SCO

Table Appendix A-1 Case Histories Sumarized by Computer Telephony Application

The following table summarizes all the case histories by end user industry.

Case History Summary By Industry				
Case Page #	Company	Computer Telephony Application	Industry	UNIX
162	IBM USA	Call Center	Retail	AIX
127	V-Systems	Fax Server	Aviation & Insurance	AIX, HP/UX, Sun
117	MediaSoft	IVR	Banking	SCO
107	IBM Canada	IVR - Student Registration	Education	AIX
255	Genesys	Call Center - Middleware	Financial	SUN
145	V-Systems	Fax Server	Financial	DG/UX
77	Voicetek	IVR	Help Desk Software	SCO & Sun
233	Arbinet	PC Switch	International CallBack	SCO
241	AlphaNet	PC Switch	Intranet based Voice transport	SCO
137	Faximum Software	Fax Sever	Manufacturing	SCO
221	Cygnus	PC Switch	Marine Ship to Shore	QNX
89	Parlance Corporation	IVR - Voice Recognition	Software	SGI IRIX
99	CallStream	IVR	Transportation	SCO
171	Teloquent	Call Center - Virtual	Utility	SCO

Table Appendix A-2 Case History Summarized by End User Industry

The following table summarizes all the case histories by type of UNIX platform. In fairness to Sun Microsystems the readership is reminded that the response to the Minden Biotech Request for Proposal by Capri Systems is based upon Sun Microsystems solutions and partners, and is a major contribution to the overall information within the book.

Case History Summary By UNIX Platform

Case Page #	Company	Computer Telephony Application	Industry	UNIX
162	IBM USA	Call Center	Retail	AIX
107	IBM Canada	IVR - Student Registration	Education	AIX
127	V-Systems	Fax Server	Aviation & Insurance	AIX, HP/UX, Sun
145	V-Systems	Fax Server	Financial	DG/UX
221	Cygnus	PC Switch	Marine Ship to Shore	QNX
117	MediaSoft	IVR	Banking	SCO
233	Arbinet	PC Switch	International CallBack	SCO
241	AlphaNet	PC Switch	Intranet based Voice transport	SCO
137	Faximum Software	Fax Sever	Manufacturing	SCO
99	CallStream	IVR	Transportation	SCO
171	Teloquent	Call Center - Virtual	Utility	SCO
89	Parlance Corporation	IVR - Voice Recognition	Software	SGI IRIX
255	Genesys	Call Center - Middleware	Financial	SUN
77	Voicetek	IVR	Help Desk Software	SUN & SCO

Table Appendix A-3 Case Histories Summarized by UNIX Platform

The following table consists of reported instances predictive dialing in a shortened form of a case history. This format does not have full vital statistics table associated with them. However there are important discussions that indicate the cost savings provided by this important technology.

UNIX based Predictive Dialing References

Page #	Company	CT Application	Industry	UNIX
187	Davox Corporation	Predictive Dialing	Financial	SUN

UNIX based Predictive Dialing References				
Page #	Company	CT Application	Industry	UNIX
186	EIS	Predictive Dialing	Cable Television	SCO

Table Appendix A-4 UNIX based Predictive Dialing References

The following are case histories yet to be written. This table is to indicate which vendors have deployed what kinds of UNIX based computer telephony applications. Because of editorial constraints and time, these will be reported further hopefully in the next edition. In the meanwhile, you may want to check out the Scheduled Solutions World Wide Web Site on the Internet at http://www.scheduledsolutions.on.ca for further updates on these case histories and check out other reviews of technology, and services. You can email the author at:

johnk@scheduledsolutions.on.ca for further information.

Future Case Histories

Company	CT Application	Industry	UNIX
Aspect Technologies	Call Center	Insurance	SCO
Digital Telecommunications Inc.	Call Center - Billing Systems	Telecoms	SCO
Scopus	Call Center - Help Desk/Genesys?	Banking	SUN
Faximum Software	Fax Server	Aviation	SCO
CTI Information Services	Fax Server - High Density CO	Telecoms	Sequent & Sun & SCO
T4 Systems	Fax Sever - High Density	Government	SCO
CallStream	IVR	Insurance	SUN
CallStream	IVR	Retail	SCO
Nortel - MSC	IVR	Marketing	SCO
Prima Telematic	IVR	Telephone Company	HP/UX
Voice Processing Plus	IVR - Order Entry/Order Status	Tire and Automotive	SCO
Vocalis	IVR - Speaker independent recognition for operator Assistance	Telecoms	SCO
CallStream	IVR & Internet - Dealer Referral	Software	SCO
Apex Voice Communications	IVR & Messaging	Social Services	Sun Interactive
Telecorp Systems	Predictive Dialer	Cable TV	SCO

Table Appendix A-5 Future Case Histories: A reference of other Vendors

Contributing Vendors and End-Users

A B Consultants
Paul Brobst
3234 McKinley Drive
Santa Clara, CA, USA
95015
1-408-243-2234
1-408-243-2236

AlphaNet Telecom
Al Gordon
55 St.Clair Avenue
Ste.400
Toronto, Ontario, Canada
M4V 2Y7
416 413-4400
416 413-4424
al.gordon@alphanet.net
http://www.alphanet.net

Apex Voice Communications Inc.
Elhum Vadhet
15250 Ventura Boulevard
3rd Floor
Sherman Oaks, CA, USA
91403
1-818-379-8400
1-818-379-8410
sales@apexvoice.com
http://www.apexvoice.com

Aspect Telecommunications
Deb Kieger
1730 Fox Drive
San Jose, Ca, USA
95131-2312
deb_kieger@aspect.com
http://www.aspect.com

Aurora
Sandy Sheer
33 Nagog Park
Acton, MA, USA
1720
1-508-263-4141
1-508-635-9756
sale@fastcall.com

BeSTpeech Products Inc.
Vince Azzara
2250 Sixth Street
Berkeley, CA, USA
94710
1-510-841-5083
1-510-841-5093
webmaster@best.com
http://www.bestpeech.com

Contributing Vendors and End-Users

Brandon Interscience Inc.
Paul Brandon
22336 Palm Avenue
Cupertino, Ca, USA
95014
1-408-257-3434
1-408-257-3443
brandon@surf.com

Brite Voice
Bill Davis
1325 Capital Street
Ste. 109
Carrollton, TX, USA
75006
1-214-323-3000
1-972-323-3010

BrookTrout
Andy O'Brien
144 Gould Street
Needham, MA, USA
2194
1-617-449-4100
1-617-449-3171
aob@brooktrout.com
http://www.brooktrout.com

Capri Systems Inc.
Steve Dunne
3031 Tisch Way, Suite 15PE
San Jose, CA, USA
95128
1-408-345-9045
1-415-851-5929
steve.dunne@internetmci.com

CTI Information Services,Inc.
Joe Enrico
11490 Commerce Park Drive
Ste. 200
Reston, VA, USA
22091
1-703-648-1610
1-703-648-1678
jenrico@ctiis.com
http://www.ctiis.com

Cygnus Technology Limited
Glen Smith
115 Main Street
Fredricton, NB, Canada
E3A 1C6
1-506-444-0696
1-506-444-0698
gsmith@cygnus.nb.ca
http://www.cygnus.nb.ca

DAC Systems
Mark Nickson
60 Todd Road
Shelton, CT, USA
6484
1-203-924-7000
1-203-944-1618

Dataquest
Chris Thompson
251 River Oaks Parkway
San Jose, CA, USA
95134
1-408-468-8632
1-408-468-8042
chris.thompson@dataquest.com
http://www.dataquest.com

Contributing Vendors and End-Users

Davox Corp.
Doug Smith
6 Technology Park Drive
Westford, MA, USA
1886
1-800-480-2299
1-508-952-0201
dougs@davox.com

Dianatel
Dan Zumar
96 Bonaventura Drive
San Jose, Ca, USA
95134
1-408-428-1000
1-408-433-3388
dtsales@dianatel.com
http://www.dianatel.com

Dimond Communications Group
Steven Fielding
264 Beacon Street
2nd Floor
Boston, MA, USA
2116
1-617-424-8373
1-617-424-1132
adimond@dimondgroup.com

EIS International, Inc.
Amy Martin
1351 Washington Boulevard
Stamford, CT, USA
6902
1-203-351-4800
1-203-961-8632
http://www.eisi.com

Genesys Labs
Karine Hagan
1155 Market Street
11th floor
San Francisco, CA, USA
94103
1-415-437-1100
1-415-437-1260
karine@genesyslab.com
http://www.genesyslab.com

Headlands Mortgage
Matthew Rapaport
700 Larkspur Landing Circle
Ste. 250
Larkspur, CA, USA
94939
1-800-462-2700
1-415-461-2733
matthew.rapaport@headmort.com
http://www.headmort.com

Contributing Vendors and End-Users

IBM Canada
Julia Klein
4175 Fourteenth Avenue
Markham, Ontario, Canada
L3R 5R5
905-316-4090
905-316-2165

IBM USA
Steve Cawn
3039 Cornwallis Road
Research Triangle Park
Raleigh, NC, USA
27709
1-919-254-7452
1-919-254-4913
http://www.raleigh.ibm.com/cti/c
tiover.html

Kowloon Canton Railway
Corporation
Percy Yeung
011 852 2688 1325
011 852 2688 1325
ppkyeung@kcrc.com

Mediasoft
Bachir Halimi
8600, Boulevard Decarie
Ste. 215
Mont-royal, Quebec, Canada
H4P 2N2
1-514-731-3838
1-514-731-3833
mst@altitude.cam.org
http://www.cam.org/*mst

Microlog
Sarah Saunders
20270 Goldenrod Lane
Germantown, Maryland, USA
20876
1-301-428-9100
1-301-540-5557
http://www.mlog.com

Natural Microsystems
Michael Katz
8 Erie Drive
Natick, MA, USA
1760
1-508-650-1300
1-508-650-1351
michael_katz@nmss.com
http://www.nms.com

Noble Systems Corporation
Bill Meaken
4151 Ashford Dunwoody Road
Ste. 550
Atlanta, Georgia, USA
30319
1-404-851-1131
1-404-851-1421
http://www.noblesys.com

Parlance Corporation
Jack Reilly
200 Boston Avenue
Medford, MA, USA
1-617-873-4636
1-617-873-2473

Power Station Technologies Inc.
Cheryl Smith
61 Pleasant Street
Randolph, MA, USA
2368
1-617-961-4400
1-617-961-4508
cheryl.smith@powerstationteck.com

Prima Telematic
Pierre Lemoine
14 Commerce Place
Ste. 510
Nun's Island, Quebec, Canada
H3E 1T5
1-514-768-1000
1-514-768-7680
pierrel@prima.com

Scopus
Stephanie D'Orazio
1900 Powell Street
Ste. 700
Emeryville, Ca, USA
94608
1-510-597-5980
1-510-597-5994
sdorazio@scopus.com
http://www.scopus.com

Arbinet
Sheila Peterson
226 East 54th Street
New York, New York, USA
10022
1-212-230-1200   x105
1-212-230-1216
allmail@dssnet.com
http://www.4smartnet.com

Stingray Boat Company
Robbie Gilbert
625 Railroad Avenue
P.O.Box669
Hartsville, South Carolina, USA
29551
1-803-383-4507   x120
1-803-332-8003
rgilbert@infoave.net
http://www.stingraypb.com/boats

Sunsoft
John Phillips
2550 Garcia Avenue
Mountain View, CA, USA
94043
1-415-786-4913
1-415-961-6078
john.phillips@eng.sun.com
http://www.sun.com

T4 Systems
Alodia Hankins
3 Innwood Circle
Ste. 116
Little Rock, AR, USA
72211
1-501-227-6637
1-501-227-6245
t4system@ix.netcom.com
http://www.t4.com

Teloquent
Jerry Geckhter
4 Federal Street
Billerica, MA, USA
1821
1-508-663-7570
1-508-663-7543
info@teloquent.com
http://www.teloquent.com

Universal Weather & Aviation
Fred Rogers
8787 Tallyho
Houston, TX, USA
77061
1-713-944-1622
1-713-943-4610
frogers@univ-wea.com
http://www.univ-wea.com

Univoice
Dave Manjarrez
147 Morgan Drive
Norwood, MA, USA
2062
1-617-255-5400
1-617-255-1980
drm@inea.usa.com

Uniworks
Bob Van Denend
922 Broad Street
Durham, NC, USA
1-919-286-1586
1-919-286-3199
info@uniworks.com
http://www.uniworks.com

Voice Processing Corporation
Daniel Dietlin
1 Main Street
Cambridge, MA, USA
2142
1-617-494-0100
1-617-494-4970
ddietlin@vpro.com
http://www.vpro.com

Voice Processing Plus
Dave Albright
5745 West Maple Road
Ste. 218
West Bloomfield, MI, USA
48322
1-810-737-9550
1-810-737-9558
vpp@quark.gmi.edu
http://www.vpplus.com

Voiceteck
Helen Chan
19 Alpha Road
Chelmsford, MA, USA
1824
1-508-250-7906
1-508-250-9378
hchan@voicetek.com
http://www.viocetek.com

Tina Stewart
Segment Marketing
Telecommunications
425 Encinal Street
Santa Cruz, CA 95061-1900
direct # 408-427-7264
fax #408-427-5411
tina@sco.com
http://www.sco.com/solutions/telephon
y/index.shtml

Appendix B

System Sizing Tables

The purpose of this Appendix is to give you a rule of thumb guide to simple lookup tool to size the number of telephone lines (ports) you may need for your computer telephony application. Please beware, these tables are a rough estimate, and by no means 100 per cent accurate for absolutely every situation. The calculations are based on a formula of 1 telephone call in every 100 would have a chance (read probability) of receiving a busy signal. There are much more detailed statistical tables that provide a more accurate answer. For further assistance you may want to contact to your telephone interconnect or telephone professional in your organization for further help. These tables are "best estimates" and other factors may need to be reviewed for better accuracy for you particular implementation.

The formula[65] for the calculation is:

# of Ports = Length of call in minutes X Number of calls per Hour X 0.0238

The following tables show increments at 1.5 minutes and 5 minutes. I've given these timings since it takes about 90 seconds to do a simple interactive voice response application, and about 5 minutes for the average conversation on a line. I have also limited the number to 128 ports, which is about the upper limit in a single computer system with multiple T1 or E1 lines terminating into a system at the time of writing. However the density of the number of ports in a single slot in a computer is increasing each year. These systems are usually networked together so that a number of units will act and behave like one large system.

Of interest is the argument by the telephone companies that dial-up modem analog Internet usage has caused them to invest heavily into new telephone switching equipment at their central offices.

The reason for this is that they have used an average of 5 minutes for a telephone conversation. Most Internet connect times are much longer and they can last for hours. This was not the original intent of the switching equipment when it was installed.

Average Length of Call: 1.50 Minutes	
# of Calls/Hour	Number of Lines
50	1.79
100	3.57
150	5.36
200	7.14
250	8.93
300	10.71
350	12.50
400	14.28
450	16.07
500	17.85
550	19.64
600	21.42
650	23.21
700	24.99
750	26.78
800	28.56
850	30.35
900	32.13
950	33.92
1000	35.70
1050	37.49
1100	39.27
1150	41.06
1200	42.84
1250	44.63
1300	46.41

Average Length of Call: 1.50 Minutes	
# of Calls/Hour	Number of Lines
1350	48.20
1400	49.98
1450	51.77
1500	53.55
1550	55.34
1600	57.12
1650	58.91
1700	60.69
1750	62.48
1800	64.26
1850	66.05
1900	67.83
1950	69.62
2000	71.40
2050	73.19
2100	74.97
2150	76.76
2200	78.54
2250	80.33
2300	82.11
2350	83.90
2400	85.68
2450	87.47
2500	89.25
2550	91.04
2600	92.82
2650	94.61
2700	96.39
2750	98.18
2800	99.96
2850	101.75
2900	103.53
2950	105.32

Average Length of Call: 1.50 Minutes	
# of Calls/Hour	Number of Lines
3000	107.10
3050	108.89
3100	110.67
3150	112.46
3200	114.24
3250	116.03
3300	117.81
3350	119.60
3400	121.38
3450	123.17
3500	124.95
3550	126.74
3600	128.52

Average Length of Call: 5.00 minutes	
# of Calls/Hour	Number of Lines
50	5.95
100	11.90
150	17.85
200	23.80
250	29.75
300	35.70
350	41.65
400	47.60
450	53.55
500	59.50
550	65.45
600	71.40
650	77.35

Average Length of Call: 5.00 minutes	
# of Calls/Hour	Number of Lines
700	83.30
750	89.25
800	95.20
850	101.15
900	107.10
950	113.05
1000	119.00
1050	124.95
1100	130.90

---

[65] Client Server Computer Telephony: The Definitive Roadmap to the Client-Server Revolution, P 12-17, by Edwin Margulies, Flatiron Publishing Inc. 12 West 21 Street, New York, NY 10010 ISBN 0-936648-55-4.

Appendix C

Intermediate Level Discussion on Database & SQL

This discussion on SQL has been written for the reader who wants to learn more about the SQL technology. The discussion continues from page 52.

The discussion is here of matter of interest, and shows how the concepts of "intersection tables" allow you to manage complex "many-to-one" relationships. This is meant for those who yearn to learn more.

The premise is based upon a student-professor-class relationships of an university.

For example a Professor table would consist of the following: (Please note that the table has been split in half in order to fit it onto the page. Normally columns for **CITY, PROV. COUNTRY, SALARY,** AND **PHONE NUMBER** would be on the right of the column **STREET**.)

Professor ID	Last Name	First Name	Street
8654125	Morgenson	Donald	65 William Street
8668316	Bezner	Hart	96 Sigma Street

City	Prov	Country	Salary	Phone Number
Kitchener	Ontario	Canada	95,000.00	519-555-8753
St. Jacobs	Ontario	Canada	95,000.00	519-555-9887

To list the Professor First Name and Last Name for a particular Last_Name: statement would be:

SELECT Last_Name, First_Name from Professor,
WHERE Last_Name="Morgenson"

Last_Name	First_Name
Morgenson	Donald

Figure Appendix C-1 Database: University Example: Professor Table, SQL Statement for Professor Search, and Results

The Course Table would look like:

Course ID	Description	Building	Room	Day	Time	Term	Length in Hours
1001	Intro to Computers	Sigmund Samuel	2-223	Tues	13:00	Fall 1996	1.5
1002	Intro to Psychology	Willison	4-222	Thurs	10:00	Winter 1997	1.0
1003	Russian History	Sigmund Samuel	2-223	Mon	08:30	Fall 1996	2.0

To list all courses scheduled for a specific room in a University building in chronological order, the SQL statement would be:

SELECT Course_ID, Description, Day, Time, Length_Hours from Course
WHERE Building = "Sigmund Samuel" **AND** Room = "2-223",
ORDER by Day, Time

The results would be:

Course ID	Description	Day	Time	Length in Hours
1003	Russian History	Monday	08:30	2.0
1001	Intro to Computers	Tuesday	13:00	1.5

Figure Appendix C-2 Database: University Example: Course Table, SQL
Statement to select a building and classroom, and show the Results

As the database administrator, you need two other special tables that builds a permanent association or relate all three tables to answer the question posed above. This type of table is called an intersection table in database terminology. This table can be generated by the relational database software as part of the process to generate the report on an *ad hoc* basis, or permanently present as part of the database. In this case it would be a permanent table, so as to minimize the compute and disk access time for such a common and frequent request. The issue is now how does one relate these different individual students, and teachers, and courses to answer our question.

The Student Schedule Table would look like:

Student ID	Course ID
78911373	1001
78911373	1002
200637311	1001
200637311	1002
200637311	1003

Figure Appendix C-3 Database:
University Example: Student
Schedule Table

Professor Schedule Table would look like this:

Professor ID	Course ID
8654125	1001
8668316	1002
8654125	1001
8668316	1002

Figure Appendix C-4 Database: University
Example: Professor Schedule Table

So looking at the Student table one can find Joanna Dobson, and find her student ID number 200637311. Then look for her student number 200637311 the Student Schedule intersection table and look up the Course time, location, and duration in the Course Table using the Course ID. Then using the Course ID in the Student Schedule table you can look up the Professor ID in the Professor schedule, and relate it back to the Professor table. This can report which professor is teaching her classes. In order to perform this the following SQL statement would be used:

```
SELECT STUDENT.last_name, STUDENT.first_name,
STUDENT.student_id,
COURSE.course_id, COURSE.description, COURSE.building,
COURSE.room COURSE.day, COURSE.time
FROM STUDENT, COURSE, STUDENT_SCHEDULE
WHERE
STUDENT_SCHEDULE.student_id=STUDENT.student_id, AND
STUDENT_SCHEDULE.course_id=COURSE.course_id AND
STUDENT.last_name="Dobson" AND
COURSE.course_term="1996 Fall"
ORDER BY COURSE.day, COURSE.time
```

Figure Appendix C-5 Database: University Example: Complex SQL Select
statement with Joins and Intersection Table Usage. See Figure Appendix C-7343
for the results.

Please do not be too concerned that the above SQL query may be as clear as mud to you. We will break it down into its component pieces so that you will get an idea of how it is working on the data in the database tables. Please note that the columns being manipulated are qualified by the name of the table to avoid confusion. This is standard SQL syntax as described by the American National Standards Institute (ANSI) SQL specification. The WHERE clause performs most of the work here in order to sort out what is going to be retrieved. First of all it takes the string:

**STUDENT.last_name="Dobson" AND
COURSE.course_term="1996 Fall"**

Figure Appendix C-6 Database: University Example: Compound Equijoins

---

And reduces the list to a student called "Dobson", that is taking courses in "1996 Fall". What is next being done is to use the intersection table STUDENT_SCHEDULE to provide the relationship between STUDENT and COURSE. In order to do this the WHERE statement does a join. It joins the two tables to produce a new set of results as a new temporary table. In this case there are effectively 2 joins happening at the same time.

The first join is:

STUDENT_SCHEDULE.student_id=
STUDENT.student_id

finds all occurrences of the same student_id that belongs to student "Dobson" in the Student table, and matches them with the same student_id in the STUDENT_SCHEDULE table.

The second join is:

STUDENT_SCHEDULE.course_id=COURSE.course_id

resolves the reference of the course_id in both the STUDENT_SCHEDULE and COURSE table by finding a course_id in the STUDENT_SCHEDULE that are the same as in the COURSE table. This narrows down the list even more.

For clarity of reporting:

ORDER BY COURSE.day, COURSE.time

This will force the output to sort the results by day of the week and the time of day of each course.

Hence the results of this query would be:

STUDENT Last Name	STUDENT First Name	STUDENT ID	COURSE ID
Dobson	Joanna	20063731	1003
Dobson	Joanna	20063731	1002
Dobson	Joanna	20063731	1002

COURSE Description	COURSE Building	COURSE Room	COURSE Day	COURSE Time
Russian History	Sigmund Samuel	2-223	Monday	08:30
Intro to Computers	Sigmund Samuel	2-223	Tuesday	13:00
Intro to Psychology	Willison	4-222	Thursday	10:00

Figure Appendix C-7 Database: University Example: Results of Complex SQL Query to list all courses for a particular student. To see the SQL statement see

Figure Appendix C-5 on page 341

Let's hark back to the beginning of this appendix, where we spoke about the Student Registration interactive voice response system. When the student called in, the caller-id would be matched in the database. Recognizing that it was a unique phone number in the Student table, it would eliminate the prompt for the Student ID, and ask for the password. Once password is verified against a security table not shown above, then it would ask for which courses she wanted to register, one at a time. She would enter the Course ID. It would be verified that it exists in the Course table, and program logic would be checked to see if she was allowed to take this course, and it did not conflict in terms of date and times with one she is already scheduled to take. Then the Student Schedule table would be updated, adding her Student ID, and Course ID. This then is how the student registration system would work in essence. After going through the registration process, and if there was an option to fax the result of a query described above, then the appropriate SQL statement would be executed, and then results would be then sent. A real student registration system would be more complex, in terms of rules and access. We will examine a case history of a major university's student registration system later in this chapter.

Let us then focus on the two published numbers by the university. They have the registration line, and the library line. The calls for both applications are terminated on the voice processing card within the computer system. The calls are hitting the voice processing card in a circular hunt group, allowing the maximum number of tries of getting an open telephone line. The call is answered on one of the ports. The first thing the interactive voice response software will do will examine the dialed-number. If dialed-number is the phone number for the student registration application, the interactive voice response software will then execute the registration application. Conversely, if the dialed-number is the number for the library application, then the interactive voice response software will then run the library application. Then within each application, the interactive voice response software could examine the caller-id, and take advantage of that to eliminate any unnecessary prompts. The result of using the dialed-number means the system administrator of the interactive voice response system does not have to dedicate telephone ports on interactive voice response system for specific applications. This means an overall reduction on the number of ports for the interactive voice response system, and as well a reduction in computer hardware, and software, and most importantly a reduction of the number of telephone lines the system requires. After all, the largest potential and ongoing cost for any interactive voice response system is the cost of the telephone phone charges and the manpower to maintain the interactive voice response application software and system. These cost money even when the system is not being used. Therefore there is emphasis in designing these large systems to reduce the number of prompts within the application (especially if the organization is paying for the line charges, unlike 1-900 services which the caller pays for the call and incurs additional charges), and to maximize of caller-id and dialed-number services from the telephone company to reduce costs.

The Fictitious Minden Biotech Request for Proposal[66]

TO PROVIDE

Multiple Site Computer Telephony
Services for Support of Customers and
Internal users
For

Minden Biotech, Inc.
A fictitious company

*Responses to this RFP demonstrate
How products are integrated into an
Enterprise Solution*

July 2 1997

[66] A very special thanks to Paul Brandon of Brandon Interscience Inc. of Cupertino, California for the framework of this request for proposal. He graciously provided special permission to reproduce portions of this request for proposal from an original real life request from a different industry with the additions made by Scheduled Solutions Inc. to change the scope, and content of the RFP. Portions of this RFP are Copyright 1997 John Kincaide All rights reserved.. John Kincaide can be contacted at Scheduled Solutions Inc., Toronto, Canada. Portions of this RFP are Copyright 1996 Brandon Interscience Inc. Used with special permission.

TABLE OF CONTENTS

Introduction

This Request for Proposal (RFP) is being issued by Minden Biotech, Inc. for proposals to engineer, furnish and install new systems for voice mail with integrated messaging, interactive voice response (IVR) and facsimile server, a call center solution including screen pops, and screen transfers, with a predictive dialing solution to solicit new potential business, and to provide follow up of test results. Currently, Minden Biotech, with headquarters in Minden, Nevada has acquired new businesses via merger and acquisition of other companies in Farmington, New Jersey, as well as in London, United Kingdom. The communication and customer service solutions within the United States are being harmonized as a major goal of this RFP. One of the goals of the RFP is to provide a common integrated solution set for both United States locations to build upon. The goals are to provide extended hours of service, by taking advantage of the time zone differences between both sites within the United States. Because of regulatory differences, the UK office can only report results within the UK for multinational corporations based in the United States. The same solution set for the London location is to be installed at the same time as the US based operations. Linking between all sites is a common communication platform, which will include long distance services, as well as distributed integrated messaging and voice mail. Since an international TCP/ip Intranet is being established for three sites, the company will also take under consideration any voice over the Internet or Intranet solutions. This will make use of the existing cost structures of the multiple T1 and E1 based Intranet. Failing an Intranet telephony solution, the company will consider an international call back system for the London office, if deemed appropriate and cost effective. The provision of a PBX and ACD technology for all three sites is requested to evaluate the cost-effectiveness of acquiring enhanced services and capabilities that this may provide.

Minden Biotech recognizes that there would be functional enhancements and efficiencies provided by a new PBX, but feels that the current older Siemens Saturn IIE systems at the Minden, Nevada office will not perform adequately for the foreseen new demands placed upon it, and that a replacement system is justified.

The company would also consider non-traditional PBX and ACD systems that take advantage of UNIX based PC Switch (UNIX workstation platforms also acceptable), in order to provide standards based and open architecture, network ready system. The Farmington, NJ office has a NEC 2200 PBX. The London, UK office has a Harris 20-20 PBX. As part of the overhaul of the communications infrastructure, the company wishes to harmonize the PBX infrastructure using PC Switch technology if possible, and/or "middleware" software technologies for the legacy PBX systems. The former rather than the latter is more preferable. The voice mail and interactive voice response systems shall be replacements of existing systems. The voice mail system will become a corporate standard and an integral method of communicating internally over 9 time zones. The company wishes to purchase an integrated UNIX voice, fax, and email messaging system that stands as an application that is independent, but communicates efficiently of the existing legacy PBX systems, as well as the newer PC Switch systems. It is the sincere desire to have a distributed preferably open database (i.e. Informix), as a storage medium for the voice, fax, and email, allowing for distributed access of the information from regular telephone & fax machines, and as well as via Netscape 3.0 or later browser Java application. This integrated messaging system would allow for messages to be retrieved via the Internet, as well as internally within the corporate Intranet.

For callers within the United States, the interactive voice response system will allow for results to be spoken and if necessary, faxed to any number within the Continental United States, Alaska, and Hawaii. In order to cut down some of the transmission costs, Minden Biotech has implemented a successful trial of email transmission of results using Internet based Pretty Good Privacy (PGP) encryption technology to selected accounts. The platform used during the trial had an OS/2 operating system with dialup UUCP access, and the company wishes to move that to a UNIX platform using C2 operating security features. This system will communicate to the Internet via a separate UNIX based Firewall system. The connection to the Internet is being upgraded to a T1 or ISDN (2B+D), where appropriate. Access to the system will be

24 hours 7 days a week. The company expects this configuration will be replicated at the different locations over time. The new interactive voice response system will allow the callers to transmit the results by PGP encrypted email to their previously authenticated email address on file within the Informix database. The authentication process to set up the accounts is controlled by a shift supervisor in the call center. These calls can not be processed by regular customer service representatives. Results cannot be transmitted to any area code in Canada, and the Caribbean. There is no formal workgroup or call center with any computer telephony integration at any of the locations. For the United States offices, the reporting function has been mainly a PC based local area network running a LAN based database. The UK office has 3rd generation language based COBOL application that runs on an SCO XENIX PC with "dumb" Wyse 60 terminals. At all locations customers are placed in a manual hold queue, and await for the next available customer agent. This database, and its application is being migrated to a new Informix Online 7.1 application that has been developed within the company. This database will run on the SCO UnixWare for the UK office on fault tolerant dual Pentium Pro system, and Sun in the New Jersey office, and IBM RS/6000 for the Nevada office. The UK office is replacing the "dumb" terminals with Windows 95 and/or Windows NT workstations. The decision to run the application on different platforms leverages existing system administration expertise in these locations.

The immediate plans are to retool and train each individual office as a stand alone computer telephony integrated call center by the end of November 1997. Each location will continue acting as a single business unit with its own reporting center. They will use the new communication technologies and Informix database applications in a localized call center solution. Integrated messaging services will be online between all locations. This comprises Phase I of the immediate project. Intensive training and staffing will occur between this time and the start of Phase II. Phase II of the project will occur starting mid February 1998. The US offices will then implement a multiple site virtual call center with skills based routing. This will allow extended hours of operation.

The facsimile server system shall replace current AT&T EasyLink service that has been used at the Nevada office. This should be UNIX based server technology, which can take advantage of time zone changes to least cost route the delivery of fax. It should have X-Window, Microsoft, and preferably Internet browser Java based technology as workstation clients. The company has multiple locations and legacy systems that are being upgraded or replaced by this RFP. There are different kinds of computer systems and desktop devices (X-Terminals within the Laboratories in Minden) as well as PC systems. There is a heightened need to utilize standards based technology such as Netscape browser.

The current voice mail system is inadequate in size and features to meet current business needs at the Minden, Nevada site. The IVR system is inadequate in feature capabilities and is expected to be undersized as the IVR service continues to grow. These deficiencies are having significant service and financial impacts on Minden Biotech. Projected cost savings achievable with the provision of an in-house facsimile server capability are significant. Consequently, it is very important to Minden Biotech that these new capabilities be provided as soon as possible.

A major desire of Minden Biotech is to acquire new systems that provide modern features and capabilities that are integrated, so that new capabilities and services can be achieved. These desires include integrated office messaging, fax confirmation option for IVR and caller transfer from IVR to an ACD group, with the capability for agent information screens (pop-screens) providing caller account information and current IVR session activity.

In summary, this RFP and its implementation represents the company's strategic move in the marketplace. Management has taken the long view that although short term Return on Investment to the shareholders is important, it is not necessarily the over riding factor in the decision making process. Management views this as a long term process that with excellent customer service, and broadening of the kinds of services to be performed in the United States market,will increase market share and overall profitability. Management sees this as an opportunity to obtain a sustainable competitive edge in the market, allowing for newer products and services to be added upon the infrastructure developed with this proposal.

Background

Minden Biotech Inc. is a leading biotech laboratory providing development of genetic management through patent recombinant DNA processes and testing services. The company has two divisions for the creation of new hybrid grains and feeds, such as wheat, barley, rice, oats and other similar products. The company also holds several patents on specialized DNA analysis quickly identifying new strains of fungal, viral, and bacteriological agents that can quickly destroy crops. Quick identification allows for farmers and agricultural agencies to take quick action with recommended procedures to combat crop losses. Remedies can include recommendations to use Minden Biotech genetically engineered strains of grains which are inherently resistant to the attack of the bacterial, or fungal agents.

Minden Biotech is certified by several U.S. and United Kingdom agencies to conduct genetic DNA work for agricultural use. Minden Biotech tests for a great number of bacterial, viral, fungal agents as provided in soil, and plant samples. In addition to genetic testing, Minden Biotech offers a range of integrated services that are customized to assist clients in implementing cost-effective genetically based testing programs.

The essential elements of genetic testing are a secure and specially handled biohazardous material container. A second independent test is performed to confirm each positive test result. The Company carefully controls each step of the testing process by following detailed written procedures, and by using the specific genetic testing methods. The Company performs the largest portion of its testing at its laboratory in Minden, Nevada that generally operates six days per week, 24 hours a day. The Company also provides complete testing services in the UK.

The Farmington[67], New Jersey office performs tests for paramedical insurance testing, which is described below. The Company has recently been awarded several patents for biotechnology testing for humans that greatly speed the results, and increase the ease and comfort for individuals providing samples for paramedical insurance testing. Management's goal in purchasing the Farmington laboratory is to eventually blend both kinds of services. This will allow the company to secure larger and more profitable customers on both US coasts, while leveraging its core biotechnology intellectual property rights. One of the objectives of this RFP is to allow customers, no matter where they may be within the US, to call one single number and be able to obtain the information they are entitled to in a quick professional and courteous manner

The steps currently taken at the Minden, Nevada, and similarly at the London, UK office to perform genetics testing process are as follows:

Specimen Collection and Transportation. Genetic testing begins with specimen collection conducted under carefully controlled conditions. Once the soil, and plant samples are collected it is assigned a unique specimen number. A bar-coded or numbered label with this specimen identification number is affixed to the specimen bottle and the bottle is sealed with a tamper-evident seal. Information pertinent to the specimen is then input onto a form that has been bar-coded or numbered to match the specimen biohazard container. The collector then prepares the specimen for shipment to Minden Biotech. Specimens, together with the forms, are delivered to the Company by overnight or same day by special courier.

- All samples taken for the Farmington, New Jersey location are shipped by special medical courier, since they contain blood samples. Shipments in the United Kingdom are also processed by special medical courier.
- Receiving: Minden Biotech receives specimens where they are inspected for tampering and checked for proper

documentation. The unique specimen identification number is entered into the laboratory computer system which automatically orders the proper screening and confirmation testing and directs the reporting of test results. A small portion of the soil sample is processed in a specialized suspension as a first step in extracting existing DNA strains in the sample. The plant portion of the sample is then forwarded to a different part of the laboratory, so its different DNA strains can be identified.

Reporting of Results. Minden Biotech transmits an increasing number of its test results electronically. Minden Biotech transmits each test result from the laboratory computer to the client's personal computer or secure facsimile machine as soon as it has been released by the certifying scientist. Using this capability, the Company routinely reports results for specimens that screen negative within 24 hours of receipt in the laboratory and within 48 hours for specimens that require confirmation. The Company also provides its clients with a telecommunications tool called Lab Line, an automated voice reporting system for transmitting drug test results to clients. Lab Line allows clients 24-hour access to certified test results via a toll-free number. Other clients receive test results via overnight courier or mail or PGP encrypted email.

Minden Biotech operations in Minden, Nevada provides genetic testing services to more than 1,000 clients in various major clients. Groups such as universities and research labs, US Military, and US Department of Agriculture, grain seed companies, farmers, and State and Local Agriculture agencies. The London, UK operation provides the same kind of services to the United Kingdom market to over 250 client in the private, and public and law enforcement agencies. With the merger the London location, the company can now provide multinational companies (i.e., transportation) that have operations in both countries with comprehensive genetic testing services.

As an adjunct business, Minden Biotech has also through the acquisition of the Farmington, New Jersey office, is also engaged in insurance medical laboratory testing. This business is somewhat similar to the kinds of testing done at the Minden head office, and in London, UK office. The purpose of testing is to approve life and health insurance coverage within the United States. The major difference is in the client base, and the multitude of different kinds of underwriter limits and specific test procedures. Each insurance company will have a different set of testing requirements for the same amount of insurance for the same life that is being insured. The testing results go to 3rd party insurance paramedical reporting agencies who in turn disseminate the information to insurance agents who originate the request. Close communication with these 3rd party reporting agencies are a very important part of the business. Besides the test results, medical examination, and specific insurance reporting forms are forwarded to the underwriter and medical staff at the insurance company by these 3rd party reporting agencies. The company is striving to automate and report back test results in the most efficient, confidential, and professional possible means.

The demographics of the Company are the following:

1. Minden, Nevada: The organization consists of 237 people. The breakdown is as follows: 65% (177) of the staff are involved in the receiving, data processing, and analysis and reporting of the samples within the laboratory. There are currently 15 customer service agents that work together to help clients with drug reports. This group operates from 07:00 Hours to 21:00 hours The MIS group consists of 23 key staff members, headed by a Director of MIS. Their key role is the maintenance and development of the new corporate Informix applications for all locations. The application is based upon an established "off the shelf" product InGenetic from the company of the same name. (*The name of the company and product are fake.*)Customization is being performed by a team consisting of the Database Administrator, Senior Systems Network and Systems administrator, a InGenetic representative, and Head of Software Engineering. Currently, Tina S. McDougal is the telephony specialist, with experience with larger call center

solutions. The respondents to this RFP must be made aware, this individual is part of a team. The team reports to the Director of MIS. The Director of MIS will make recommendations to the management of the company to approve the purchases of products. Last but not least, the remaining staff members consist of the President, CFO, Operations VP, the administration, and accounting.

2. Farmington, New Jersey has 150 people involved in the business of providing laboratory testing of urine, and blood for insurance medical examinations. Although the number of 3rd party reporting companies are about 16 major clients, the Farmington laboratory processes a very high volume of results, as compared to the Minden, Nevada, and London, UK locations. Each major account will forward over 1000 samples a week, resulting, on average about 16,000 results. There will be aggressive ongoing sales and marketing campaigns in the western United States. This is expected to increase business at a rate of 25% to 35% per annum for next 2 years in the laboratory and call center activities. Computer telephony integration and Internet services are key parts of the infrastructure to handle this growth. This business unit is managed by a team of a Vice President responsible to the overall profitability and Operations Manager for the laboratory. There also exists a MIS team of 5 individuals who maintain the Sun based systems.

3. London, United Kingdom office has 70 people. This too is headed by a Vice President, who reports to the President in Minden, Nevada. This organization has been a bit more autonomously managed than the other operating locations. It also has its own chief financial officer for financial reporting needs that are particular to the UK way of doing business. The demographics are much the same as the Minden Nevada location, in that about 60% of the staff are involved in the receiving of samples and the analysis, and reporting of the same. The new Informix applications and network tools will greatly increase this portion of the company's ability to respond much quicker to the newer market opportunities. Of significant importance is the recent passing of legislation in Parliament to allow firms to

provide genetic testing in other European Community markets in October 1998. The automation and call center capabilities, it is hoped will allow the company to communicate more effectively with their customers. Although this market is more regulated than in the United States, the need to respond to newer market opportunities is even greater in the UK.

---

[67] This place name used in this RFP does not reflect reality, as best known to the author at the time of writing. The company in this RFP is fictional. Any name or person, place or thing that matches a real entity is purely coincidental and unintentional.

Current Systems and Planned Expansion

CURRENT SYSTEMS

The current voice mail system is a Acme VMail 5000. This system is connected to the PBX with 8 ports and provides voice storage capacity of 7 hours. The primary problem with the system is that the processor is unable to keep up with the load. The results of this are that there may be before an incoming message is posted to a user's account and calls are dropped in busy hours.

The current IVR system is connected to the PBX with 8 ports. This system is to be replaced by an integrated UNIX based product, that will allow for easy software and hardware expandability by Minden Biotech staff. The product should come in the following preference with a graphical user interface based upon a Netscape Java application, X-Window, or alternatively a Microsoft Windows application. The product should have full Informix Online 7.1 (with an upgrade path to Informix Universal Server in the near future) access and support from both the UNIX servers. The IVR reporting capability will be available for all 3 sites. As part of Phase I, the vendor will provide and install a copy of the software and hardware at all 3 locations. Upon the start of Phase II of the project, the IVR systems within the United States are to be "harmonized," allowing for calls to overflow from each site. At the beginning of Phase II, the two US locations will offer and support both types of laboratory services. The Informix database will have been centralized, so applications will dynamically be loaded based upon the Dialed Number Information Service (DNIS) and correspondingly the database on the network will be used to report results. The London, UK system will for now stand alone. However, it must be able to be perform similar types of networked database and possibly telephony based access. This will allow the company to provide remote access to the Continental market in the autumn of 1998.

The PBX is a Siemens Saturn IIE single cabinet system with 4 remaining usable card slots before an expansion cabinet must be added. The pertinent information on current and planned new

connectivity is provided in Table Appendix D-6 Current and
Planned PBX System Capacities at Minden, Nevada on page 364.

The Company wishes the respondents to find an alternative
solution to the standard PBX technology. The current trends in
the marketplace to replace these systems with "PC Switches" and
"PC as Phone" technologies will be considered first, before
standard legacy PBX technology. The tables included within the
RFP state the number of and types of telephone sets, that an
equivalent number of computer users with "PC as Phone"
technology should be used under a site license, or active user
license.

PLANNED PBX EXPANSION

Minden Biotech will be responsible for ordering all of the telco
trunk expansion reflected in the table on the following page.
Minden Biotech will also be responsible for all of the PBX port
expansion reflected in the table on the following page. This will
include additional quantities reflected in the "New Quantity"
column of the table and changes reflected in the "Comments"
column. These additions and changes will be implemented by
Minden Biotech before cut over of the systems requested in this
RFP.

Current Quantity	New Quantity	Description	Comments
235	251	Analog line ports equipped or equivalent in the PC Switch.	Sets, modems, fax, IVR, voice mail, announcers, etc. (Sufficient spare capacity for all of the "New Quantity" addition of ports)
158	158	Telephone sets or PC as Phone user licenses.(A preference for a "Floating user license" to reduce costs.)	139 analog, 19 digital (DYAD-18 sets). If a PC as Phone solution is offered as part of a PC Switch, then the PC Switch should also support 12 analog lines for regular analog phones (i.e. Plain Old Telephone Set) support. These set will be

Current Quantity	New Quantity	Description	Comments
			placed in the reception area, and each major working area to allow visitors, and emergency personnel access to a regular phone. PC as Phone software will work through the TCP/ip backbone of the corporation. (see note 1)
8	8	Voice mail PBX ports	
	8	Auto-Attendant PBX ports	Auto Attendant function for incoming calls on main numbers (702-555-5200 and 800-555-4177) (see note 1)
8	12	IVR PBX ports	4 new ports for anticipated growth (see note 1)
	4	Office Fax PBX ports	New ports for incoming office Fax (from DID) and outgoing office Fax (see note 1)
	10	Fax Server direct (not through the PBX)	Outbound for Fax delivery of drug test results
22	22	Pacific Bell trunks	main number (702-555-5200) incoming and local outgoing
8	10	Pacific Bell DID trunks	702-555-82xx (2 new ports for adding office fax mailbox capacity, also 2 new 100 number DID groups to be added for more direct numbers and fax mailbox numbers)
48	48	Sprint trunks (2 - T1)	45 voice (incoming and outgoing LD) and 3 data. Voice includes: 800-555-4177 (main 800 number) 800-555-0048 (reports desk agents) 800-555-7378 (IVR results information)
8	8	Call Center ACD agent	Covers 702-555-5200 and 800-555-4177

Current Quantity	New Quantity	Description	Comments
		positions	
7	7	Reports Desk ACD agent positions	Covers 800-555-0048

Table Appendix D-6 Current and Planned PBX System Capacities at Minden, Nevada

Note 1: Voice port "New Quantities" shown for Voice Mail, Auto Attendant, IVR and Office Fax represent what is necessary for service on the existing PBX which uses in-band signaling for transfer information and cannot reflect the trunk group source. This requires separate port groups for action unique to the source type. For Proposals including a PBX replacement with better transfer information, some of these port groups may be combined depending upon the configuration proposed. If all of these services for Voice Mail, Auto Attendant, IVR and Office Fax are provided on one platform, the total ports to be provided between the PBX and the server platform shall be 20. This may be provided as a T1 link if that option is available and it is deemed to be cost-effective by the Vendor.

Current Quantity	New Quantity	Description	Comments
157	192	Analog line ports equipped or equivalent in the PC Switch.	Sets, modems, fax, IVR, voice mail, announcers, etc. (sufficient spare capacity for all of the "New Quantity" addition of ports)
147	162	Telephone sets or PC as Phone user licenses (A preference for a "Floating user license" to reduce costs.)	15 "Digital Display Sets", a single attendant, and remaining analog sets.
4	8	Voice mail PBX ports	
4	8	Auto-Attendant PBX ports	Auto Attendant function for incoming calls on main numbers (206-555-4200 and 800-555-3867) See Note 2 below about phone numbers and their usage at Phase II.
	4	IVR PBX ports	4 new ports for anticipated growth are to be added at the beginning of Phase II in 1998.
	2	Office Fax PBX ports	New ports for incoming office Fax (from DID) and outgoing office Fax (see note 1)
	12	Fax Server direct (not through the PBX)	Outbound for Fax delivery of drug test results
12	15	New York Telephone trunks	main number (201-555-4200) incoming and local outgoing
8	10	New York Telephone DID trunks	201-555-72xx (2 new ports for adding office fax mailbox capacity, also 1 new 100 number DID groups to be added for more direct numbers and fax mailbox numbers)
24	24	Sprint trunks (1 - T1)	24 voice (incoming and outgoing LD). Voice

Current Quantity	New Quantity	Description	Comments
			includes: 800-555-4167 (main 800 number) 800-555-0058 (reports desk agents) 800-555-7358 (IVR results information)
8	8	Call Center ACD agent positions for Reports Desk.	Covers 201-555-5200 and 800-555-4177

Table Appendix D-7 Current and Planned PBX System Capacities for Farmington, New Jersey

Note 2: When Phase II of the project is implemented, there will be a harmonization of all the 800 phone numbers among the two US locations. Customers will be given one all encompassing phone number that the marketing will publish. This will then allow callers on the autoattendent either the person internally, or the department, or Reports Desk or Lab Line. Callers for the Reports Desk, or callers who have requested an agent from the Lab Line IVR will be answered by any agent in **both** sites. So the ACD queues have to monitor activity in both sides of the country, and as well any those agents that are remote off site customer service agents.

Current Quantity	New Quantity	Description	Comments
72	96	Analog line ports equipped or equivalent in the PC Switch.	Sets, modems, fax, IVR, voice mail, announcers, etc. (sufficient spare capacity for all of the "New Quantity" addition of ports). Expectations are that this switch will need to handle larger loads in 1998, when deregulation of this service occurs. Economical quick modular expansion is a concern.
	70	Telephone sets or PC as Phone user licenses (A preference for a "Floating user license" to reduce costs)	6 digital display sets, an attendant set, and the remaining are analog sets. If a PC as Phone solution is recommended, then a floating user license would be of interest. Floating is defined here as the number of actual potential outbound calls that can be achieved by the switch. If there are 32 available lines (i.e. E1), then a license for 32 will be issued. However, this does not prevent any number of users to have the software up and running on their workstation, and making requests of the PC Switch or PBX.
	4	Voice mail PBX ports	Currently there is no Voice mail at the London location.
	4	Auto-Attendant PBX ports	Auto Attendant function for incoming calls on main number 0165 223 615
	2	IVR PBX ports	4 new ports for anticipated growth in 1998
	2	Office Fax PBX ports	New ports for incoming office Fax (from DID) and outgoing office Fax .

Current Quantity	New Quantity	Description	Comments
	2	Fax Server direct (not through the PBX)	Outbound for Fax delivery of drug test results, but this is to double in October 1988.
1	30	British Telecom E1	main number (0165 223 615) incoming and local outgoing
4	7	Reports Desk ACD agent positions	

Table Appendix D-8 Current and Planned PBX System Capacities at London, United Kingdom

Besides the voice communications, the company is installing as part of their data communications network among all three sites the following high speed circuits for database, Intranet, services.

Minden, Nevada: two T1s

Farmington, Nevada 1 T1.

London, UK 1 E1.

VENDOR QUESTIONS AND MEETINGS

Any questions that the Vendor may have shall be addressed directly to Tina McDougal at Minden Biotech will meet with each Vendor individually, if desired by the Vendor, to allow an opportunity for the Vendor to review the facility. Meetings must be scheduled with at least 24 hours advance warning.

Proposal Requirements

The Vendor shall submit a proposal conforming to all of the following requirements to be considered for evaluation.

PARTIAL PROPOSAL RESPONSES

Vendors are invited to submit partial Proposal responses for portions of this RFP. Partial responses shall be acceptable for:

Phase I: Individual office site Implementation

1. **PBX system replacement at all sites. A single vendor will be selected to supply all sites. Again, the company will favorably review a UNIX based "PC Switch" systems.**
2. **Office Call Processing** (including voice mail and fax mail, integrated desktop messaging, fax from desktop applications, auto-attendant and IVR)
3. **Intranet or Internet long distance telephony services** capabilities between sites, in order to cut down internal traffic costs. If this service is available for outside the company use, with excellent sound quality, the company would be interested to

THE FICTITIOUS MINDEN BIOTECH REQUEST FOR PROPOSAL

an respondent who can supply such services, or be brought in house as part of the UNIX communications technology infrastructure.

4. **Fax Server -** This product provides a vital service to transmit test results to the clients. It also should provide multiplatform support in either client-server mode, or native platform. It also should also be incorporated, if possible, with the Fax Mail component of the Integrated Messaging software.

5. **Call Center - Application Software:** This could be the "middleware" to a recommended legacy private branch exchange or PC Switch component that will allow the efficient transferring of calls to the next available agent . The InGenetic application will be "screen popped". Details as to the number of agents that will be available will be located in the office.

The Company will also accept solutions that will allow for ISDN based remote call center staff. The main concern is security of the information. However, the Company has a stated policy for telecommuting. If the employee can demonstrate that a "secure" portion of the home is possible to their manager, then it is to the advantage of both parties that this can be implemented. Skills based routing will be based upon skills such as the level of expertise, and genetic knowledge of the results, and as well as seniority. Unlike other call centers, the staff at Minden Biotech are highly trained professionals, who understand the biohazard issues and implications. Database rules based programming will be even more evident and important at the Farmington, New Jersey office, since InGenetic software will be encoding the paramedical rules for each kind of test. (Different kinds of testing from different life insurance companies for the same life health status being insured. There are many rules, and this is to be integrated for the IVR, Predictive Dialer, as well as for the Call Center software at Farmington.)

1. **Predictive Dialer -** One of the key marketing features will be the Company's ability to perform outbound calls to our clients with notification of test results. Notification is to be done only to a verified

person, not to any recording device. Positive results (i.e., presence of biohazard) will be relayed as soon as possible after the verification of the test results. This will allow our staff to place a more human touch to our customer's needs. Optionally, the customer's may also want their Negative results also performed over the phone. This function can be performed within the offices or via telecommuters, and should allow for blending of both inbound and outbound calls. The Predictive Dialer will use the standard InGenetic screens. A new marketing move by the Farmington office is to provide the major accounts with Positive results (i.e. insurance medical failure, is presence of illegal drugs, or nicotine or other pharmacological agents that would invalidate a health or life insurance policy.)

PHASE II: INTERCOMMUNICATION BETWEEN SITES:

1. For the United States offices: The ability for IVR, and call center applications to see the results of both major lines of results (Genetic and Insurance paramedical) to be handled across the country. The company can then offer services for all 50 States (i.e., Hawaii, and Alaska included).
2. That office Integrated Messaging can now be shared between all locations. Voice, fax, and Email can be forward, courtesy copied, to any one location. Sales and other off site staff can also receive this information.
3. PC Switch and/or PC as Phone technology, can allow any one member of the company to organize and use their "PC as Phone software" to create conference calls easily, and take advantage of the switch features via drop down menus, and via easy to use of setup scripts (i.e. "Wizards"), that were hitherto buried within the myriad of arcane touch-tone codes of legacy proprietary PBX phone.
4. If implemented, a Intranet based telephony service for use within the corporation in order to reduce costs.

5. Skill based routing and blending of inbound calls and predictive dialing between both call center sites within the United States sites to cover 6 time zones from Hawaii to the eastern United States.

Vendors are invited to submit Proposals for any or all of these components. Vendors should realize that all major centralized components of their response should be residing on Open Systems UNIX based servers. The structure of the RFP is such that the best products from all UNIX platforms will be considered. If one vendor platform can not fulfill all the needs on the same UNIX platform, then the Company may select another UNIX platform offering for that component. Although the UNIX servers supporting the InGenetic software are on different UNIX platforms, it is not necessarily the bell weather platform for the UNIX computer telephony solutions at that location. Other UNIX solutions can be recommended. The Company will publish the results as an open bid, indicating what products were selected and for what cost.

PROPOSAL FORMAT

The Proposal shall include an item by item response for each numbered section and sub-section of this RFP. All sub-sections in RFP sections 1 through 6 may be satisfied by the response "Read and understood." Other sub-sections may be responded to in the same way if the sub-section only provides information for the Vendor. Any sub-section that requires equipment or action by the Vendor shall have a narrative response describing how the vendor shall meet the requirement. If the Vendor is not submitting a Proposal that includes a specific section, the response for that section shall be "Not being proposed."

The Vendor may include any reasonable length cover letter to introduce the Proposal, discuss strengths, highlight approach, etc. that the Vendor may feel important. The Proposal shall include appendices or attachments of pertinent equipment, software, programming and user manuals. The Proposal may also include additional information in appendices or attachments that the

372

Vendor feels is useful for Minden Biotech to have for review and evaluation.

The RFP is intentionally brief and succinct in feature and functional descriptions to allow for flexibility to accommodate differing product capabilities and Vendor approaches. It is therefore very important that the Vendor provide sufficient descriptive information covering all of the proposed equipment, software and services offered in order for the Proposal to be considered. All Proposals and proposed equipment and functions will be carefully and thoroughly evaluated. Sufficient information for this evaluation must be provided for the Proposal to be considered.

PROPOSAL DUE DATE

Two copies of the proposal shall be delivered to Minden Biotech, Inc., 149 Douglas County Airport Drive, Minden, Nevada, no later than 5:00 p.m. on Friday August 1, 1997.

Proposal Evaluation Criteria

The Proposals will be evaluated based upon three primary areas of concern:

1. **Functional capability in meeting the needs required by Minden Biotech in this RFP.** Some consideration will be given to features and capabilities available in the Vendor offerings beyond the specifics of the RFP requirements that Minden Biotech feels may be useful, either now or in the future.

2. **Financial evaluation of the Proposal.** This will include the price to Minden Biotech of the Vendor offerings as well as consideration of cost impacts or savings that may accrue to Minden Biotech in the use of the Vendor offerings.

3. **Schedule for cut over of the systems offered.** It is of significant importance to Minden Biotech to have the systems in and operational as soon as possible. Minden Biotech is experiencing customer service impacts due to the non-performance of current systems. Additionally, there are cost savings immediately available to Minden Biotech upon completion of systems cut over. Since it is expected that no one single provider will be available solve or provide all components, the company would recommend that as part of Phase I, that a "proof of concept" test of integrated technology be presented, as part of the overall cost of the proposal. This would allow for the vendors to proof out the products, and allow for development of better deployment for the final implementation of Phase I in the autumn of 1997.

4. **Integration and installation of products offered.** Integration of products into a modularized and open standard based communications architecture is an evaluation priority.

5. **Summation of ROI for the company, and Minden Biotech' customer ROI (i.e. "Ripple Effect").** This can be stated in terms of dollars saved or earned, over a stated period of time. See the "Vital Statistics" table at the end of the RFP to tabulate the benefits. Replicate or expand as you need.

6. **Training, and "Triage Teams" Support.** Vendors are to detail the training and costs for the system administration, and end users of the software being proposed. The training is not only centered on the pre and post delivery of the equipment and software at each Phase of the project. The respondents are asked to provide any ongoing training services that they can provide on a quarterly basis, in order to help our staff to maximize the usage of the equipment. Each staff member is allocated a minimum of 1.5 to 2 days per month towards training. The Company values its employees, and regular training every quarter is scheduled to increase the human wealth of the corporation. The Company is very interested in developing "Triage Teams" that can respond 24hours a day to any systems crisis in United States as well as in London. Vendors are to provide training and support, and advanced early detection and monitoring technologies to eliminate downtime when it occurs. If technologies like software systems that provide Simple Network Management Protocol (SNMP) solutions to remotely monitor, diagnose, and resolve downtime issues are available, they would be appropriate.

PBX Replacement

The vendor shall provide a new modern PBX system for replacement of the current Siemens Saturn IIE at the Minden, Nevada location. This replacement should preferably be a UNIX based PC Switch solution, which instead of regular phones, PC as Phone software will replace the regular handsets. (Please note, the

Company is planning to purchase PCs or PC Workstations or NC workstation machines as they become available.) A TCP/ip network is the basis of all communication as much as possible. If that is not part of the vendor solution, then computer telephony vendors should recommend a modern legacy PBX model and manufacturer they would recommend to interface, using "middleware" technologies. Again if the PBX can be part of the regular TCP/ip network, the more favorable the solution would be to the company. The recommended interface would be via this middle layer, so that the actual telephone PBX hardware and software are insulated from the proprietary nature of these systems. The intent would be to allow for a PBX replacement to another manufacturer in the future, without a lot of down time and hassle to retrain and reprogram the software to work with it. The Vendor shall provide detailed information on the system, attendant and set features for evaluation.

ANALOG SET SUPPORT FOR EXISTING SETS

The new system shall provide service for the existing analog telephone sets, only if no computer based PC as Phone software technology can be used for the general office. The call centers are to have headsets in place for ease of use These include:

Quantity	Set Type	Comments
73	Siemens Euroset 2212	With message waiting light
13	Siemens Euroset 2102	With message waiting light
12	Vodavi 2604E	With message waiting light
29	Miscellaneous analog feature button sets	

DISPLAY SETS

The current display sets are Siemens DYAD-18 display sets with 18 feature buttons. These sets are used for secretarial call coverage (4) and ACD agent positions (14). The new display sets shall provide equivalent functionality and a similar quantity of programmable feature/line buttons.

ACD Service

ACD Service is very important to Minden Biotech. The current need is for 2 ACD groups. Additional ACD groups are envisioned in the near future. The new system must be capable of supporting up to 8 ACD groups. The agent and call statistics reporting and display capability of the new system must be comprehensive. The Vendor shall provide detailed information with the Proposal describing these capabilities for evaluation.

Please be aware that Farmington, and London require a minimum on 1 ACD queue each for the call centers, with growth possibilities similar to that of Minden, Nevada.

Cabling

The new system shall utilize the existing wiring, cabling and distribution system as much as is reasonable. Any new wiring, cabling or distribution needed to support the new system including connection of the system monitor, remote access modem connection, connection to all trunks and peripheral equipment, etc. shall be provided by the Vendor as a part of the system installation. This will include any distribution systems additions or changes needed to accommodate the system installation location selected by the Vendor in consultation with Minden Biotech.

System Reliability

Good system reliability is necessary. The Vendor shall provide information on MTBF and MTTR expected for the system configuration proposed. Battery backup is required. The following specific reliability features are required:

1. The Vendor shall provide reasonable redundancy of power supplies.
2. The Vendor shall provide redundant CPU

   3. Capability for rapid system backup and reload of software shall be provided.

   4. Modem access and/or Internet Telnet, and FTP access if supported, for remote maintenance shall be provided.

SYSTEM CAPACITIES

The system capacities shall be configured to meet all necessary set, trunk and peripheral equipment interface needs for all systems detailed in this RFP. The ultimate capacity of the new system shall be at least twice the wired capacity specified in Table 7-1. New system capacities are presented in Table 7-1. The Vendor shall utilize this table as a guide, but shall be responsible for proposing any additional elements necessary to meet the total configuration needs for the systems proposed by the Vendor

Equipped Quantity	Wired Quantity	Description	Comments
240	256	Analog line ports	Sets, modems, fax, IVR, voice mail, announcers, etc.
19	24	Digital display set ports	Call coverage (4) and ACD (15)
1	1	Attendant Console	
4		18 button Digital Display sets	Call coverage positions
15		18 button Digital Display sets	ACD agent positions
4	4	T1 digital trunks	2 - Sprint 800 + LD trunks 2 - Pacific Bell trunks 22 - main number in + local out 10 - DID 16 - OPX

Figure Appendix D-8 New PBX System Capacities for Minden, Nevada location

Please be aware that the Farmington location will need about 50% of capacities of Figure Appendix D-8.

Enterprise wide Voice Mail and Office FAX

The Vendor shall furnish an office voice mail and office fax capability, to provide individual subscriber voice and fax mailboxes and voice announcement mailboxes. This important messaging system will allow users between all major locations, and remote locations, access via regular telephone and fax (and email by fax), as well as through preferably platform independent Netscape browser application. If possible, an open database technology that can be manipulated like another database application in a client/server environment would be desirable.

1. If the Proposal does not include a PBX replacement, the system shall interface with the current Minden Biotech Siemens Saturn IIE PBX.

2. The system platform shall be provided with a hot-swap disk capability through RAID disk striping level 5 for assured protection of all disk information.

The proposed system shall be compatible with three types of PBX interface protocols including:

- Inband DTMF
- Serial RS232 Data
- Simplified Message Desk Interface (SMDI)
- TCP/ip based messaging

BASIC VOICE MAIL FEATURES

The office voice and fax mail capability shall include at least the following features and capabilities:

1. Personal Greetings
2. Pager Notification
3. Fax Reception
4. Message Forwarding
5. Direct Message Reply (on-system subscriber)
6. Direct Call Return (on-system subscriber)
7. Message Time & Date Stamp
8. Message Fast Forward/Rewind and Pause
9. Urgent , Private, Regular, Message confirmation, and scheduled delivery Options
10. Group Messaging
11. Recovery of deleted messages during and after access of mailbox.
12. Off Premise Delivery of messages with a mailbox owner ability to define phone numbers and schedule of delivery.

GENERAL SYSTEM STATISTICS

The system shall provide for general system administrative statistical reports and display. These statistics shall include at least the following information with appropriate separate statistics for voice and fax:

1. Number of calls per port per day
2. Total calls per hour
3. Total call minutes per hour
4. Number of messages sent per hour
5. Number of mailbox accesses per hour
6. Total message storage used and total capacity

INTEGRATED MESSAGE SOFTWARE

Voice and Fax Messaging

The system shall include integrated messaging software to allow for desktop access to all messages from a single application program. The software shall integrate message information display and access for:

1. Voice messages -- if the resources are available, messages can be played back and responded to by using either

   a) If a multimedia PC or Workstation with speakers are present, with a microphone for both internal users and external remote users

   b) If remotely connected via a slower analog line, a user can designate a regular telephone as an outbound dialing function to listen and respond.

2. Fax messages

3. Integrated TCP/ip Simple Mail Transmission Protocol software support for UNIX mail access and remote POP3 mail support as well.

E-Mail Messages

It is desirable for the integrated messaging software to include e-mail as well as voice and fax messages. The e-mail interface desired is for Internet e-mail. It is anticipated that Minden Biotech will implement a full-time Minden Biotech Internet node for business customer access of test result information and e-mail use. The Vendor shall describe how this need for integration of e-mail messages in the integrated messaging software shall be met in the near future and when it will be available, if it is not now.

DESKTOP FAX CAPABILITY

Support for centralized facsimile capability to support office desktop PC fax needs shall be provided. This shall allow for the preparation and transmission of faxes from individual desktop computers on the network.

1. File Format Support
   a) ASCII text
   b) PCL 4 & 5 (HP LaserJet Files)
   c) PostScript
   d) Any file printable within Windows
2. Transmit features:
   a) Automatic cover sheet generation
   b) Distribution list processing (Broadcast Fax)
   c) Programmable retry strategy
   d) Off-hours scheduling
   e) Multiple site least cost routing of faxes, and fax store forward technology to maximize costs
   f) Standard and fine mode resolutions
   g) Automatic retransmission of interrupted fax pages
3. Management features:
   a) Priority scheduling levels
   b) Detailed transmit logs
   c) Real time status reporting
   d) Customizable header
4. Overlay capabilities:
   a) Forms
   b) Logos
   c) Signatures

CAPACITIES AND INTERFACES

The capacities and interfaces to be supported by the proposed equipment and software are:

1. 8 port voice mail interface
2. 24 hours of voice mail storage capacity
3. 18 CCS[68] traffic capacity for office fax (in and out)
4. Storage capacity for 10,000 fax pages

5. Storage capacity for 20,000 pages of ASCII e-mail and attachments
6. TCP/ip network Interface
7. 32 bit Integrated Message software for Netscape 3.0 or later browser for all display platforms
8. Microsoft Windows 32 bit application will be considered as an alternative. However, since not all platforms the company currently owns are PC based, the company does has a preference for a Java based Netscape application that is platform independen.

---

[68] CCS is "Centi Call Seconds. One hundred call seconds or one hundred seconds of telephone conversation. One hour of telephone traffic is equal to 36 (60 x 60 = 3600 divided by 100 = 36) which is equal to one erlang. CCS is used in network optimatization...." Quotation taken from Newton's Telecom Dictionary page 115, Copyright 1996 Harry Newton, Published by Flatiron Publishing, 12 West 21 Street, New York, New York 10010.

Automated Attendant

The system shall provide the capability for automated attendant answering on specified primary number groups with customized scripts. Numbers to be supported shall include the main listed number, the main 800 number (DNIS) and specific published 800 information numbers.

The proposed system shall be compatible with three types of PBX interface protocols including:

- Inband DTMF
- Serial RS232 Data
- Simplified Message Desk Interface (SMDI)
- TCP/ip based messaging.

SPECIFIC CAPABILITIES AND FEATURES

The required capabilities and features include:

1. Office hours greeting
2. After hours greeting
3. Holiday greeting
4. Receptionist mailbox greeting
5. Connect a caller to a telephone number
   a) internal
   b) external
1. Receive caller input
   a) via touch-tone
   b) via speech recognition
2. Play back information to a caller
   a) via recorded speech
   b) via text-to-speech conversion

AUTOMATED ATTENDANT STATISTICS

Automated Attendant statistics are required for management and administration.

Statistics Information

Statistics shall include coverage for all automated attendant activity, as well as pertinent statistics by number group for the following types of information:

1. Number of calls per day
2. Total calls per hour
3. Average length of call per hour
4. Total call minutes per hour
5. Number of messages received per hour
6. Number of extensions dialed
7. Number of extensions answered
8. Number of extensions not answered and caller left a message
9. Number of extensions not answered and caller pressed "0" for assistance
10. Number of extensions not answered and caller tried another extension
11. Number of calls abandoned

Statistics File Output

The statistics shall be displayed on the system administrative console, printed or exported to files. The export files shall output in forms suitable for inclusion in spreadsheet, database or word processing programs. The Vendor shall specify the available formats for file export.

CAPACITIES

The Automated Attendant shall support main listed numbers (702-555-5200 and 800-555-4177) in a single automated attendant group. Additionally, 5 specific published 800 information numbers shall be supported, each with separate automated attendant routines. The capacities required for these are:

1. Main listed number group at 60 CCS
2. Information numbers at 45 CCS total for all 5 numbers

Total storage of 3 hours shall be provided for recorded messages (to caller) and message mailboxes (from caller) for these automated attendant calls. At the initiation of Phase II, a single 1-800 toll free line will be issued to simplify access for customers. An autoattendendent will direct callers to the appropriate IVR, and ACD queues for either plant or human test results.

PLATFORM RELIABILITY

The platform supporting the Automated Attendant function shall have mission-critical hardware reliability. The system platform shall be provided with a hot-swap disk capability through RAID disk striping level 5 for assured protection of all disk information.

Interactive Voice Response (IVR)

The IVR system is used for customer call-in and receipt of test results information. This information will be retrieved from a Minden Biotech Informix Online Dynamic Server 7.1 database. Two forms of information shall be retrieved from the database; customer profile data and test results data. Additionally, the IVR process must update the test results record on the Informix Online Dynamic Server 7.1 database to reflect the last time the record was reported to the customer through IVR.

SPECIFIC IVR SYSTEM REQUIREMENTS

Specific elements required of the IVR system include:

1. The IVR system shall readily interface with a Informix Online Dynamic Server 7.1 database system.
2. The customer profile data shall be used by the IVR to select or control elements of the choices and responses provided for the caller in accordance with unique characteristics for each customer.
3. The IVR shall include the capability for the caller to select an option to be transferred to a Reports Desk ACD agent.
4. The IVR shall include the capability for touch-tone input or speaker independent speech recognition to be used.
5. The IVR shall include the capability for the caller to select an option to have the results information just listened to, sent to the caller by fax to a pre-defined (in the customer profile from the database) fax number.
6. The proposed system shall be compatible with three types of PBX interface protocols including:
   - Inband DTMF
   - Serial RS232 Data
   - Simplified Message Desk Interface (SMDI)

REPORTING STATISTICS

Statistics Information

The system shall maintain usage statistics for management and administrative purposes that shall include:

1. Number of calls per port per day
2. Total calls per hour
3. Total call minutes per hour
4. Average call/minutes for all ports within a given range of dates (one day or several days).

Statistics File Output

The statistics shall be displayed on the system administrative console, printed or exported to files. The export files shall output in forms suitable for inclusion in spreadsheet, database or word processing programs. The Vendor shall specify the available formats for file export. Using OBDC or Informix based products to an Informix database tables would be preferable.

IVR SYSTEM PROGRAMMING & CALL CENTER SCREEN POPS & PREDICTIVE DIALING

The system shall provide an easy to use programming language for configuration of the IVR functions. The Vendor shall provide detailed information with the Proposal on the programming language and commands used in the IVR system.

The configuration programming for the IVR function shall be performed by Minden Biotech and the Vendor is not required to include configuration programming activity for the IVR in the Proposal.

SYSTEM CAPACITY

The IVR system in Minden, Nevada shall support a single listed number (702-555-7378) for the results reporting function. The capacity required for this operation is for 225 calls per hour peak at 90 seconds average duration for 203 CCS of traffic. The

Farmington IVR is expected to have 300 CCS of traffic, while the London UK system, accepting calls only from calls originating in the UK until October 1, 1998, shall have about 100 CCS of traffic. During Phase II the callers within the US will be able to call one number and results will be provided no matter if it is a genetic , or insurance paramedical laboratory result request.

PLATFORM RELIABILITY

The platform supporting the IVR function shall have mission-critical hardware reliability. The system platform shall be provided with a hot-swap disk capability through RAID disk striping level 5 for assured protection of all disk information.

REPORTS DESK POP-SCREEN

Minden BiotechMinden Biotech needs "Pop-Screen" capability for the InGenetic Reports Desk ACD agents to have account information available on the screen that is unique to the account of the calling party. Integration between the PBX must be achieved using either:

a) "Middleware" software technology residing on a UNIX platform, which would signal both the InGenetic application screen delivery to an agent, as well as the ACD function of the legacy PBX. TCP/ip and database table accesses, will be the means of communication between the middleware UNIX platform to the Informix based InGenetic application.

b) If the PBX is not used, but a PC Switch technology is used then communications must be open and non proprietary, and easily modifiable by system administration. TCP/ip and database table accesses will be the means of communication between the middleware UNIX platform to the Informix based InGenetic application.

The three situations that may be developed for this Pop-Screen capability are:

1. Direct incoming calls for the Reports Desk agents to a voice prompt for a caller ID number input. This information would be used to activate a specific InGenetic routine to query the Minden Biotech Informix Online Dynamic Server 7.1 database for customer account information to display on the agent screen when the call was directed to that agent.

2. A variation for caller transfers from the IVR function would be to also include information on the activities and results just experienced by the caller on the current IVR interaction.

3. The ability for customers who have been provided security clearance to review test results online via an Internet connection, to initiate a call from their Java application, and either choose to have an outbound call at a later time, (i.e. this is where predictive dialing or scheduling will come to play) to the customer, or setup an "on-demand" conference call between the customer and the call center. In both cases the designated phone number(s) is on file within the InGenetic system, with a request for authentication via telephone keypad input once the contact is made with the customer.

The Vendor shall discuss in the Proposal how the proposed IVR system, and call center applications may accommodate the development needs if undertaken in the future.

PREDICTIVE DIALER

Although the volume of outbound calls for both Minden, Nevada, and London, UK are not very large, there is a need to contact customers, if they have not received their test results by any means (i.e. via the Lab Line, Internet, or have not called in) within 24 hours of the results being completed by the laboratories. This results on average about 50 to 75 calls per shift per day, that the staff at each location need to manage. Besides these calls, the company will want to use a predictive dialer for the sales and marketing departments at all locations in order to do market surveys, as well as for the new sales campaigns especially in all regions of the business in 1998 after Phase II of the project is complete.

A special mention is required during Phase I of the project for Farmington, NJ in that their business is different than the other locations, and the need for a predictive dialer is even greater. Because of the large volume of number of test results, there is a need to process 200 to 400 calls per shift.

CALL CENTER VOLUMES

The current locations have the following agent staffing and call volumes, with 25% to 35% average growth rate per year starting late 1997 and into 1998. These call volumes reflect only inbound calls. As part of Phase II the Minden, and Farmington call centers will work in unison to handle *all* both genetic, and insurance paramedical laboratory results from a central single published toll free number. The InGenetic databases will reside in a client-server environment between locations, allowing access from any workstation (with authorization) to blending of both inbound and outbound calls

Location	Number of Agents	Average Number & Length (minutes) of Calls Per Hour	Number of Skill Levels for Skill based Routing
Minden, Nevada	15	300 calls/hr -- 7 minutes	3
Farmington, NJ	13	220 calls/hr -- 10 minutes	6
London, UK	4	100 calls/hr -- 6 minutes	3

Table Appendix D-9 Current Lab line Call Volumes for the locations.

It is important to note that the number of calls per hour and minutes are a best estimate. The company is aware that they have not been able to answer all the calls, and there is no real estimate of the number of lost calls held in a manual queue.

Facsimile Server

Minden Biotech reports test results to many customers through automated fax transmission. This is currently supported through AT&T EasyLink service. The Facsimile Server function will allow for these facsimile transmissions to be supported in-house. This need is currently for transmission of ASCII files generated by laboratory automation software. The output of this process generates batch files with header information and end-of-message lines delineating the fax transmissions. Each of these fax transmissions is a unique point-to-point fax and is not broadcast fax.

Additionally, the new Facsimile Server function will need to support informational and promotional fax transmissions to customers on a broadcast basis.

FACSIMILE SERVER CAPABILITIES

In addition to the current ASCII file and informational broadcast fax needs, it is expected that the new system will be used for enhanced format and functional capabilities in the future. The total spectrum of the needs envisioned is outlined below.

1. File Format Support
    a) ASCII text
    b) PCL 4 & 5 (HP LaserJet Files)
    c) PostScript
    d) Any file printable within Windows, or from a Java application within a Netscape browser.
2. Transmit features:
    a) Automatic cover sheet generation
    b) Distribution list processing (Broadcast Fax)
    c) Programmable retry strategy
    d) Off-hours scheduling
    e) Standard and fine mode resolutions
    f) Automatic retransmission of interrupted fax pages

394

3.   Management features:
    a)   Priority scheduling levels
    b)   Detailed transmit logs (see Section 0 below)
    c)   Real time status reporting
    d)   Customizable header

4.   Overlay capabilities:
    a)   Forms
    b)   Logos
    c)   Signatures

ADMINISTRATIVE STATISTICS AND DETAILED TRANSMIT LOG

Detailed logging of transmissions information is required. This is necessary for legal reasons as well as management information.

Administrative Statistics

The system shall maintain usage statistics for management and administrative purposes that shall include:

1. Number of calls per port per day
2. Total calls per hour
3. Total call minutes per hour

Statistics File Output

The statistics shall be displayed on the system administrative console, printed or exported to files. The export files shall output in forms suitable for inclusion in spreadsheet, database or word processing programs. The Vendor shall specify the available formats for file export.Using OBDC or Informix based products to an Informix database tables would be preferable.

Detailed Transmit Log

A detailed transmit log must be maintained on all test result transmissions. This log must be accessible for display or report generation from the Minden Biotech network. This log must contain the following information:

1. date transmission was successful
2. time transmission was successful
3. phone number
4. fax type (from file header)
5. file identifier (from file header)

In addition, it is necessary to generate an automatic exception notification for every occurrence of non-successful transmission. This non-delivery information must also be stored for display and report generation access from the Minden Biotech network.

The Vendor shall discuss in the Proposal how these needs shall be met.

PLATFORM RELIABILITY

The platform supporting the Fax Server function shall have mission-critical hardware reliability. The system platform shall be provided with a hot-swap disk capability through RAID disk striping level 5 for assured protection of all disk information.

CAPACITIES

Broadcast fax generation capability of 1 broadcast fax per day of 2,500 recipients shall be provided. Capacity shall be provided to support 325 fax transmissions per hour averaging 1.2 pages per fax. This transmission capacity includes both test results fax needs and broadcast fax needs. It is assumed that the broadcast fax load will be distributed off-peak by priority scheduling levels.

Specifications

The Vendor shall provide detailed specifications on all hardware to be provided. This shall include information such as power requirements, BTU loading, physical sizes, disk capacities, disk speeds, memory capacity, etc. as necessary to evaluate the capability of the systems and operational support needs.

Block Diagrams

The Vendor shall provide block diagrams as appropriate to illustrate all connectivity required between all systems provided by the vendor, and between Vendor provided equipment and Minden

THE FICTITIOUS MINDEN BIOTECH REQUEST
FOR PROPOSAL

Biotech provided equipment. This shall include cabling or channel quantities, speeds or sizes, etc. as necessary to fully understand the systems being proposed and all of the connectivity elements and requirements.

Installation

VENDOR RESPONSIBILITY

The Vendor shall provide and be responsible for all systems hardware and software installation and configuration. This shall include all feature and table setup and configuration necessary for a fully turn-key system (excluding IVR configuration and application programming as discussed in Section 0, and cooperative programming work between the Vendor and Minden Biotech for Fax Server Transmit Log needs as discussed in Section 0). Installation shall include the provision of all miscellaneous support apparatus and hardware including connecting cables, mounting hardware, system monitors, etc. necessary for a complete and fully functional system at the proposed total price.

MINDEN BIOTECH RESPONSIBILITY

Minden Biotech shall be responsible for the ordering and provision of all telco trunk additions and changes. Additionally, if the current PBX is retained, Minden Biotech shall be responsible for any changes or additions to the port and trunk needs for system to accommodate the new systems acquired through this RFP. Shortly after beginning work, the Vendor shall provide a detailed list of PBX and trunk changes to be provided by Minden Biotech to accommodate the Vendor provided equipment.

SCHEDULE

Because of current service problems, the installation schedule is a significant concern to Minden Biotech. The Vendor shall provide in the Proposal, an outline of the key milestone dates for installation of all systems proposed. Because of the needs for programming work by Minden Biotech on the IVR system and cooperative work in programming the Fax Server, it is important that these systems be on-site and available for programming prior to full systems operations.

Equipment and Software Warranty

The Vendor shall provide detailed information on the equipment and software warranty provided for all systems and software proposed.

Maintenance

PBX MAINTENANCE

It is required that Proposals for new PBX equipment include first year maintenance as a part of the PBX system contract. The Vendor shall specify in the Proposal the levels of service outage definitions and MTTR objectives for each outage. A copy of the proposed maintenance agreement shall be provided with the Proposal.

SERVER SYSTEMS MAINTENANCE

It is desirable that the Vendors provide first year hardware maintenance as a part of the contract for all computer equipment proposed for Office Server (Voice Mail, Auto Attendant, IVR, Office Fax and Integrated Messaging) and Fax Server (results fax delivery). The Vendor shall indicate in the Proposal whether first year hardware maintenance is included and provide the details of any proposed first year hardware maintenance.

Pricing

Pricing shall be provided for each of the system options being proposed. This pricing shall be separate pricing for each option such that Minden Biotech may elect to acquire any of the options individually from the Vendor for the price quoted without having to acquire all options.

The following tables present suggested outlines for pricing that may be used by the Vendor. These may be modified by the Vendor to meet specific Proposal needs. Minden Biotech desires to obtain as much detail as can reasonably be provided for the purpose of understanding what is being proposed and for Proposal evaluation.

OPTION 1 PRICING - PBX

Quantity	Description	Unit Price	Extended Price
1	Hardware and System Software		
1	Installation		
4	Call Coverage display phones		
15	ACD display phones		
1	End-User training		
1	Redundant hardware (CPU, power)		
1	ACD		
1	Management software		
	Total Price		
1	Second year annual maintenance		

OPTION 2 PRICING -- OFFICE CALL PROCESSING

Quantity	Description	Unit Price	Extended Price
1	Hardware		
1	Hardware Installation		
1	Voice Mail software		
1	Auto Attendant Software		
1	IVR software		
1	Office Fax software		
1	Integrated Messaging client software site license		
1	First year hardware & software maintenance		
	Total Price		
	Second year hardware maintenance		

401

OPTION 2 PRICING -- FACSIMILE SERVER

Quantity	Description	Unit Price	Extended Price
1	Hardware		
1	Hardware Installation		
1	Fax server software		
1	First year hardware maintenance		
	Total Price		
	Second year hardware & software maintenance		

CALL CENTER "SCREEN POP" WITH "MIDDLEWARE" TECHNOLOGY

Quantity	Description	Unit Price	Extended Price
1	Hardware		
1	Hardware Installation		
1	"Screen Pop" software		
1	Middleware software -- if applicable		
1	First year hardware maintenance		
	Training		
	Total Price		
	Second year hardware & software maintenance		

CALL CENTER PREDICTIVE DIALING SOFTWARE TECHNOLOGY

Quantity	Description	Unit Price	Extended Price
1	Hardware		
1	Hardware Installation		
1	Predictive Dialing		
1	Middleware software -- if applicable		
1	First year hardware maintenance		
	Training		
	Total Price		
	Second year hardware & software maintenance		

Vital Statistics

Cost	*Statistics*
Weeks of labor to install system	
Initial System Cost	
Est. Ongoing support Cost	
System Size	
# Ports - Total/# Fax.	
# Users	
# Calls/month	
Computer	
Operating System	
Industry	
Estimated Return on Investment	
Cost Savings	
Manpower equivalency per year	
Estimated Manpower savings per year	
Reduction in telephone costs/year	
Earnings	
Generated new income/year	
Ripple Effect	
Estimated cost Saving to customer base	
Vendor	
Product	
End User	
Computer Telephony Technologies Deployed	
for your solution please indicate the technology you are deploying and enter "YES" in the next column.	

Glossary

DOS A computer operating system originally developed by Microsoft Corporation for the IBM PC. The acronym means Disk Operating System. It is software that controls the computer resources and schedules other programs to execute. It was designed as a single tasking operating system, allowing one task to be operated at one time.

IVR Interactive Voice Response: One of the major Computer Telephony application classes of programs. A caller is requested to enter digits from the telephone keypad or in some instances speak phrases into the phone, in response to recorded phrases. This information is then analyzed and usually processed against a database for correct response. Then the caller is presented either with a menu of choices or the requested information can be spoken, faxed, or even emailed as a text message to the caller. An example of interactive voice response application is banking by phone, to provide balances, transaction history, and current exchange rates to a caller. The benefit of this technology is that it frees up staff from regular questions to do other work. IVR is used in conjunction with other computer telephony applications, such as call center or computer telephony workgroup applications.

LAN A Local Area Network (LAN) allows computer users in each office to share and print information amongst other users in that office. This is achieved by joining computers through a network of cabling and associated software. The software includes the protocol that is used to move the data through the network. The most common software network protocol used with UNIX is TCP/ip. The Internet networking protocol is TCP/ip based. You do not use TCP/ip itself as a user application program. It runs behind the scenes moving data between the computers on the LAN or WAN systems. You may have used an application that uses TCP/ip for networking such as a world wide web browser to 'surf' the Internet, or FTP (File Transfer Program) or Telnet for terminal emulation to different systems. Usually the LAN consists of two types of architectures. These are called server based, and Peer-to-Peer systems. The server architecture consists of one computer that is dedicated or not dedicated to providing the file and print services for all other computers that are linked to it. A UNIX system is used as a non-dedicated network server. This means that many computers can use its resources at the same time, while it can perform other tasks not related to managing the network, such as being a communications server for voice and fax, and as well as a database server. Moreover UNIX based systems can be configured to be the main server in a LAN or to have UNIX servers working together in a Peer-to-peer architecture. Peer-to-peer systems have no one system that is dominate within the LAN. All the computers in the network have the ability to share data and print services on any one computer in the network. In order to maintain the sanity of the system administrator who maintains and troubleshoots LAN problems, there will be some type of data, printing and network discipline enacted for the network for every user to follow. This is especially true with the nature of Peer-to-peer LANs.

Scalability. A computer term that describes the ability to easily scale the hardware and/or software of a computer system. This allows the owner of the system to simply add more computer hardware processing capability, or software licenses to an existing system with minimal downtime of the computer system itself.

WAN A Wide Area Network - A Wide Area Network is a system of networked computers which are usually geographically located apart from each other. For example, a company may have an office in Toronto, Canada and another office located in Los Angeles, United States of America. Each office would have a network of computers usually referred to as a Local Area Network or LAN. A LAN allows computer users in each office to share and print information amongst the computer users in that office. A WAN refers to the communications architecture that bridges two or more LANs together, so that users in any one user in one city or locale can utilize and share the computing resources and information in another.

INDEX

Notice to the readers

If you feel that there is a concept or entry, or sub-entry, that you think should be included or changed, please do not hesitate to email your suggestion to Scheduled Solutions' office at johnk@scheduledsolutions.on.ca or by other means as listed in the contact information for the company within the book. Your suggestions will be very much appreciated.

A

B

C

D

E

F

G

H

I

Example of author's company's registered domain of scheduledsolutions.on.ca · 244
Internet Protocol addressing scheme · 244
Internet Service Provider
random assignment of IP addresses to customers · 245
Internet Service Providers - Function they perform · 245
InterNIC - organization that manages IP addresses · 244
Internet Phone · *See* Voice on the Internet (VON)
ITU-T · 151. *See* United Nations
IVRitis, a.k.a Voice Mail Jail · 90, 92, 97

J

Java Telephony Application Programming Interface (JTAPI) · 208–12
JavaSoft
JTAPI · 208–12
JTAPI · 208–12

K

KCRC · *See* Kowloon Canton Railway Corporation
Kowloon Canton Railway Corporation · 99–105
Advance Warning of Pickup of cargo · 102
Bilingual issues
Cantonese & English resolved · 101
CallStream's VoiceStream IVR · 101
Integration with V-Systems' VSI-FAX · 102
Number of calls taken manually per day · 101
Wagon Information System III (WIS3) · 100

L

Linkon Corporation
Fax Servers · 156

M

Maritel · *See* WJG Maritel
MediaSoft · 158
The National Bank of Egypt · 117
Microsoft · 3, 5, 11, 407
DOS · 3, 5, 11, 407
Microsoft's NetMeeting · 243
NT · 3

Q

R

S

T

U

W

Windows Telephony Application Programing Interface · *See* TAPI under
  Computer Telephony Integration
WJG Maritel · **221**
  Area of Coverage · 223
  Cygnus's use of QNX real time based UNIX · 223

X

Xircom
  ISDN Products Division · 223

Y

York University
  Call Flow
    Access window for course registration · 111
  Digital Equipment Corporation VAX · 110
  IBM DirectTalk/6000 · 109
  Management Issues
    Concentration of Skill Sets in staff · 113
    Proactive trouble shooting by DirectTalk/6000 · 112
  Oracle
    Multithreaded database increases throughput · 112
    The advantage of using business rules within · 110
  Oracle used for Student Information System · 110
  Student Registration Interactive Voice Response System · **105–15**
  Typical manual registration before automation · 107–9
York Unversity
  background · 109

Let us help you!

Do you want Scheduled Solutions to help you create competitive advantage and enhanced shareholder value?

Scheduled solutions provides professional services to the Computer Telephony industry and as well as to corporations looking to install Computer Telephony technology.

Here is what we can do to help your company

- We provide seminars to educate corporate managers on the effectiveness of Computer Telephony Integration. This includes how to track the ROI and the Ripple Effect, how to build and develop effective RFPs, and how to manage the acquisition, and deployment of computer telephony technology. These seminars can be range from a 1 hour presentation to a half-day to full-day formats, depending on your needs. We can also provide customized courses.

- We also provide consulting services to help managers in the development of a computer telephony solution.

- Sales training seminars for Computer Telephony companies that need to have their sales staff made aware of the benefits and sales techniques to effectively sell CTI.

- We also provide sales and marketing support for Computer Telephony companies software and hardware companies. This includes providing knowledgeable and effective sales and pre-sales capabilities. This is performed on either a short term or long term contractual basis.

Company Overview

Scheduled Solutions Inc.

6 Glen Edyth Drive
Toronto, Ontario
Canada
M4V 2V7
416-929-0440 Phone
416-929-7927 Fax
johnk@scheduledsolutions.on.ca
www.scheduledsolutions.on.ca

John Kincaide, President of Scheduled Solutions Inc. founded the company in 1986. Scheduled Solutions Inc.'s mission is to provide front-line sales, marketing and pre-sales technical support for computer telephony integration software, hardware manufacturers and system integrators. The company also provides end user consulting and educational services on computer telephony integration. Scheduled Solutions has been involved in the UNIX marketplace since 1988. Since 1993 the company has been focused on computer telephony integration.

John Kincaide has developed new business for his clients in Hong Kong, the Middle East, Europe and North America. He has a solid track record in designing and marketing solutions that include interactive voice response (IVR), computer telephony integration (CTI) products on SCO UNIX and other UNIX platforms. One of John's major strengths is an acute understanding of business processes and how they relate to the technology.

Besides being the author of "UNIX Computer Telephony - The Complete Guide", John is a speaker on computer telephony, where he has addressed seminars at *SCO Forum 1995, 1996, and 1997* in Santa Cruz California and *Catalog System Management Network* conference in Memphis Tennessee, September '95, and has been quoted in *Uniforum Monthly (September '95)*, *SCO World (December '95)* , *Communication Week (December '95)*, *PC Week (October, 1996)* and *SCO World (December '96)*. John has his

Bachelors degree with a double major in history, and psychology, and computing courses from Wilfrid Laurier University, Waterloo, Ontario, Canada.

Stay in Touch!

Stay in Touch with each other and win a free Scheduled Solutions T-shirt! We want to hear from you - our readership. Scheduled Solutions wants to help share the experiences of other business managers installing computer telephony technology, through selected submissions to John Kincaide at johnk@scheduledsolutions.on.ca We will gather the most interesting discussions, and distribute them by email to those who sign up for the service. Please be assured the email list is not to be shared with others. The service is free until further notice.

To entice you to participate, we will provide readers of the book a chance to win a free Scheduled Solutions T-Shirt, just by registering by fax, or by email the following form, or go to our Web site www.scheduledsolutions.on.ca. One entry per person. T-shirts come in one size that fits all - Large, and one color - White. They have images of the "The Cobbler", or "The Panicky CEO on Holiday", or the "The Good Doctor in Montreal". There is 1 winner each month. Registration form is on page 429. The draw happens until further notice on the Scheduled Solutions Website, or by email to registered readers, and is drawn from the current readership list who have registered by the end of each month. Delivery by regular post, please allow 2 to 6 weeks for delivery.

Also check out our Website for more details about our services, and scheduled speaking engagements and seminars on computer telephony.

You can contact us by either emailing us at johnk@scheduledsolutions.on.ca, or by pointing your browser to: www.scheduledsolutions.on.ca, or by fax at 416-929-7927. We would really like to hear from you, your comments, issues, and problems.

T-Shirt Registration Form

You can register as well by copying this page (Remember Copyrights laws do exist! Photocopy this page only please!) and filling out the form below and faxing it to our office at 416-929-7927. You can also email the same information to johnk@scheduledsolutions.on.ca with the Subject Line "T-SHIRT REGISTRATION".

Do not worry, you name and information will not be distributed to others for marketing purposes, without your permission.

PLEASE PRINT:

First Name:_____

Last Name: _____

Title:_____

Company:_____

Street: _____

City: _____

Province/State:_____

Postal Code/Zip: _____

Country: _____

Business Phone Number: _____

If outside Canada or United States please include (Country code, area code/city code and phone number)

Fax Phone Number: _____

If outside Canada or United States please include (Country code, area code/city code and phone number)

Internet based or Internet accessible email address:

_____

World Wide Web address:

_____

☐ **Check here if you wish to be contacted by a representative of Scheduled Solutions Inc. at no obligation**

☐ **Check here if you allow Scheduled Solutions to inform you about products and services from other companies**

THANK-YOU!